by John McPhee

The Control of Nature

THE

CONTROL

OF NATURE

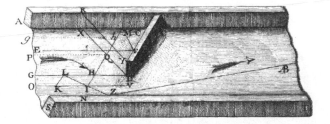

JOHN McPHEE

Farrar, Straus and Giroux

New York

Farrar, Straus and Giroux
18 West 18th Street, New York 10011

Printed in the United States of America
Published in 1989 by Farrar, Straus and Giroux
First paperback edition, 1990

Illustrations by Gudjon Olafsson from Vestmannaeyjar:
byggd og eldgos © 1973 by Gudjon Armann Eyjolfsson.
Reprinted by permission of Isafoldprentsmidja hf.

The text of this book originally appeared
in The New Yorker.

Library of Congress Cataloging-in-Publication Data
McPhee, John A.
The control of nature / John McPhee. — 1st ed.
1. *Environmental protection.* 2. *Man — Influence on nature.*
I. *Title.*

TD170.M36   1989   304.2 — dc19   89-1052

Paperback ISBN-13: 978-037-452259-9
Paperback ISBN-10: 0-374-52259-6

Designed by Cynthia Krupat

www.fsgbooks.com

37 38

For Vanessa, Katherine, Andrew, and Cole

CONTENTS

Atchafalaya

THREE HUNDRED MILES up the Mississippi River from its mouth—many parishes above New Orleans and well north of Baton Rouge—a navigation lock in the Mississippi's right bank allows ships to drop out of the river. In evident defiance of nature, they descend as much as thirty-three feet, then go off to the west or south. This, to say the least, bespeaks a rare relationship between a river and adjacent terrain—any river, anywhere, let alone the third-ranking river on earth. The adjacent terrain is Cajun country, in a geographical sense the apex of the French Acadian world, which forms a triangle in southern Louisiana, with its base the Gulf Coast from the mouth of the Mississippi almost to Texas, its two sides converging up here near the lock—and including neither New Orleans nor Baton Rouge. The people of the local parishes (Pointe Coupee Parish, Avoyelles Parish) would call this the apex of Cajun country in every possible sense—no one more emphatically than the lockmaster, on whose face one day I noticed a spreading astonishment as he watched me remove from my pocket a red bandanna.

"You are a coonass with that red handkerchief," he said.

A coonass being a Cajun, I threw him an appreciative smile. I told him that I always have a bandanna in my pocket, wherever I happen to be—in New York as in Maine or Louisiana, not to mention New Jersey (my home)—and sometimes the color is blue. He said, "Blue is the sign of a Yankee. But that red handkerchief—with that, you are pure coonass." The lockmaster wore a white hard hat above his creased and deeply tanned face, his full but not overloaded frame. The nameplate on his desk said RABALAIS.

The navigation lock is not a formal place. When I first met Rabalais, six months before, he was sitting with his staff at 10 A.M. eating homemade bread, macaroni and cheese, and a mound of rice that was concealed beneath what he called "smoked old-chicken gravy." He said, "Get yourself a plate of that." As I went somewhat heavily for the old chicken, Rabalais said to the others, "He's pure coonass. I knew it."

If I was pure coonass, I would like to know what that made Rabalais—Norris F. Rabalais, born and raised on a farm near Simmesport, in Avoyelles Parish, Louisiana. When Rabalais was a child, there was no navigation lock to lower ships from the Mississippi. The water just poured out—boats with it—and flowed on into a distributary waterscape known as Atchafalaya. In each decade since about 1860, the Atchafalaya River had drawn off more water from the Mississippi than it had in the decade before. By the late nineteen-forties, when Rabalais was in his teens, the volume approached one-third. As the Atchafalaya widened and deepened, eroding headward, offering the Mississippi an increasingly attractive alternative, it was preparing for nothing less than an absolute capture:

before long, it would take all of the Mississippi, and itself become the master stream. Rabalais said, "They used to teach us in high school that one day there was going to be structures up here to control the flow of that water, but I never dreamed I was going to be on one. Somebody way back yonder—which is dead and gone now—visualized it. We had some pretty sharp teachers."

The Mississippi River, with its sand and silt, has created most of Louisiana, and it could not have done so by remaining in one channel. If it had, southern Louisiana would be a long narrow peninsula reaching into the Gulf of Mexico. Southern Louisiana exists in its present form because the Mississippi River has jumped here and there within an arc about two hundred miles wide, like a pianist playing with one hand— frequently and radically changing course, surging over the left or the right bank to go off in utterly new directions. Always it is the river's purpose to get to the Gulf by the shortest and steepest gradient. As the mouth advances southward and the river lengthens, the gradient declines, the current slows, and sediment builds up the bed. Eventually, it builds up so much that the river spills to one side. Major shifts of that nature have tended to occur roughly once a millennium. The Mississippi's main channel of three thousand years ago is now the quiet water of Bayou Teche, which mimics the shape of the Mississippi. Along Bayou Teche, on the high ground of ancient natural levees, are Jeanerette, Breaux Bridge, Broussard, Olivier—arcuate strings of Cajun towns. Eight hundred years before the birth of Christ, the channel was captured from the east. It shifted abruptly and flowed in that direction for about a thousand years. In the second century A.D., it was captured again, and taken south, by the now unprepossessing Bayou Lafourche, which, by the year 1000, was losing its hegemony to the river's present course, through the region that would be known as Plaquemines. By the nineteen-fifties, the Mississippi River had advanced so far past New Orleans and out into the Gulf that it was about to shift again, and its offspring Atchafalaya was ready to receive it. By the route of the Atchafalaya, the distance across the delta plain was a hundred and forty-five miles—well under half the length of the route of the master stream.

For the Mississippi to make such a change was completely natural, but in the interval since the last shift Europeans had settled beside the river, a nation had developed, and the nation could not afford nature. The consequences of the Atchafalaya's conquest of the Mississippi would include but not be limited to the demise of Baton Rouge and the virtual destruction of New Orleans. With its fresh water gone, its harbor a silt bar, its economy disconnected from inland commerce, New Orleans would turn into New Gomorrah. Moreover, there were so many big industries between the two cities that at night they made the river glow like a worm. As a result of settlement patterns, this reach of the Mississippi had long been known as "the German coast," and now, with B. F. Goodrich, E. I. du Pont, Union Carbide, Reynolds Metals, Shell, Mobil, Texaco, Exxon, Monsanto, Uniroyal, Georgia-Pacific, Hydrocarbon Industries, Vulcan Materials, Nalco Chemical, Freeport Chemical, Dow Chemical, Allied Chemical, Stauffer Chemical, Hooker Chemicals, Rubicon Chemicals, American Petrofina—with an infrastructural concentration equalled in few other places—it was often called "the American Ruhr." The industries were there because of the river. They had come for its navigational convenience and its fresh water. They would not, and could not, linger beside a tidal creek. For nature to take its course was simply unthinkable. The Sixth

World War would do less damage to southern Louisiana. Nature, in this place, had become an enemy of the state.

Rabalais works for the U.S. Army Corps of Engineers. Some years ago, the Corps made a film that showed the navigation lock and a complex of associated structures built in an effort to prevent the capture of the Mississippi. The narrator said, "This nation has a large and powerful adversary. Our opponent could cause the United States to lose nearly all her seaborne commerce, to lose her standing as first among trading nations. . . . We are fighting Mother Nature. . . . It's a battle we have to fight day by day, year by year; the health of our economy depends on victory."

Rabalais was in on the action from the beginning, working as a construction inspector. Here by the site of the navigation lock was where the battle had begun. An old meander bend of the Mississippi was the conduit through which water had been escaping into the Atchafalaya. Complicating the scene, the old meander bend had also served as the mouth of the Red River. Coming in from the northwest, from Texas via Shreveport, the Red River had been a tributary of the Mississippi for a couple of thousand years—until the nineteen-forties, when the Atchafalaya captured it and drew it away. The capture of the Red increased the Atchafalaya's power as it cut down the country beside the Mississippi. On a map, these entangling watercourses had come to look like the letter "H." The Mississippi was the right-hand side. The Atchafalaya and the captured Red were the left-hand side. The crosspiece, scarcely seven miles long, was the former meander bend, which the people of the parish had long since named Old River. Sometimes enough water would pour out of the Mississippi and through Old River to quintuple the falls at Niagara. It was at Old River that the United States was going to lose its status among the world's trading nations. It was at Old River that New Orleans would be lost, Baton Rouge would be lost. At Old River, we would lose the American Ruhr. The Army's name for its operation there was Old River Control.

Rabalais gestured across the lock toward what seemed to be a pair of placid lakes separated by a trapezoidal earth dam a hundred feet high. It weighed five million tons, and it had stopped Old River. It had cut Old River in two. The severed ends were sitting there filling up with weeds. Where the Atchafalaya had entrapped the Mississippi, bigmouth bass were now in charge. The navigation lock had been dug beside this monument. The big dam, like the lock, was fitted into the mainline levee of the Mississippi. In Rabalais's pickup, we drove on the top of the dam, and drifted as well through Old River country. On this day, he said, the water on the Mississippi side was eighteen feet above sea level, while the water on the Atchafalaya side was five feet above sea level. Cattle were grazing on the slopes of the levees, and white horses with white colts, in deep-green grass. Behind the levees, the fields were flat and reached to rows of distant trees. Very early in the morning, a low fog had covered the fields. The sun, just above the horizon, was large and ruddy in the mist, rising slowly, like a hot-air balloon. This was a countryside of corn and soybeans, of grain-fed-catfish ponds, of feed stores and Kingdom Halls in crossroad towns. There were small neat cemeteries with ranks of white sarcophagi raised a foot or two aboveground, notwithstanding the protection of the levees. There were tarpapered cabins on concrete pylons, and low brick houses under planted pines. Pickups under the pines. If this was a form of battlefield, it was not unlike a great many battlefields—landscapes so quiet they belie their story. Most battlefields, though, are places where something happened once. Here it would happen indefinitely.

We went out to the Mississippi. Still indistinct in mist, it looked like a piece of the sea. Rabalais said, "That's a wide booger, right there." In the spring high water of vintage years— 1927, 1937, 1973—more than two million cubic feet of water had gone by this place in every second. Sixty-five kilotons per second. By the mouth of the inflow channel leading to the lock were rock jetties, articulated concrete mattress revetments, and other heavy defenses. Rabalais observed that this particular site was no more vulnerable than almost any other point in this reach of river that ran so close to the Atchafalaya plain. There were countless places where a breakout might occur: "It has a tendency to go through just anywheres you can call for."

Why, then, had the Mississippi not jumped the bank and long since diverted to the Atchafalaya?

"Because they're watching it close," said Rabalais. "It's under close surveillance."

AFTER THE CORPS dammed Old River, in 1963, the engineers could not just walk away, like roofers who had fixed a leak. In the early planning stages, they had considered doing that, but there were certain effects they could not overlook. The Atchafalaya, after all, was a distributary of the Mississippi—the major one, and, as it happened, the only one worth mentioning that the Corps had not already plugged. In time of thundering flood, the Atchafalaya was useful as a safety valve, to relieve a good deal of pressure and help keep New Orleans from ending up in Yucatán. The Atchafalaya was also the source of the water in the swamps and bayous of the Cajun world. It was the water supply of small cities and countless towns. Its upper reaches were surrounded by farms. The Corps was not in a political or moral position to kill the Atchafalaya. It had to feed it water. By the principles of nature, the more the Atchafalaya was given, the more it would want to take, because it was the steeper stream. The more it was given, the deeper it would make its bed. The difference in level between the Atchafalaya and the Mississippi would continue to increase, magnifying the conditions for capture. The Corps would have to deal with that. The Corps would have to build something that could give the Atchafalaya a portion of the Mississippi and at the same time prevent it from taking all. In effect, the Corps would have to build a Fort Laramie: a place where the natives could buy flour and firearms but where the gates could be closed if they attacked.

Ten miles upriver from the navigation lock, where the collective sediments were thought to be more firm, they dug into a piece of dry ground and built what appeared for a time to be an incongruous, waterless bridge. Five hundred and sixty-six feet long, it stood parallel to the Mississippi and about a thousand yards back from the water. Between its abutments were ten piers, framing eleven gates that could be lifted or dropped, opened or shut, like windows. To this structure, and through it, there soon came a new Old River—an excavated channel leading in from the Mississippi and out seven miles to the Red-Atchafalaya. The Corps was not intending to accommodate nature. Its engineers were intending to control it in space and arrest it in time. In 1950, shortly before the project began, the Atchafalaya was taking thirty per cent of the water that came down from the north to Old River. This water was known as the latitude flow, and it consisted of a little in the

Red, a lot in the Mississippi. The United States Congress, in its deliberations, decided that "the distribution of flow and sediment in the Mississippi and Atchafalaya Rivers is now in desirable proportions and should be so maintained." The Corps was thereby ordered to preserve 1950. In perpetuity, at Old River, thirty per cent of the latitude flow was to pass to the Atchafalaya.

The device that resembled a ten-pier bridge was technically a sill, or weir, and it was put on line in 1963, in an orchestrated sequence of events that flourished the art of civil engineering. The old Old River was closed. The new Old River was opened. The water, as it crossed the sill from the Mississippi's level to the Atchafalaya's, tore to white shreds in the deafening turbulence of a great new falls, from lip to basin the construction of the Corps. More or less simultaneously, the navigation lock opened its chamber. Now everything had changed and nothing had changed. Boats could still drop away from the river. The ratio of waters continued as before—this for the American Ruhr, that for the ecosystems of the Cajun swamps. Withal, there was a change of command, as the Army replaced nature.

In time, people would come to suggest that there was about these enterprises an element of hauteur. A professor of law at Tulane University, for example, would assign it third place in the annals of arrogance. His name was Oliver Houck. "The greatest arrogance was the stealing of the sun," he said. "The second-greatest arrogance is running rivers backward. The third-greatest arrogance is trying to hold the Mississippi in place. The ancient channels of the river go almost to Texas. Human beings have tried to restrict the river to one course— that's where the arrogance began." The Corps listens closely to things like that and files them in its archives. Houck had a point. Bold it was indeed to dig a fresh conduit in the very ground where one river had prepared to trap another, bolder yet to build a structure there meant to be in charge of what might happen.

Some people went further than Houck, and said that they thought the structure would fail. In 1980, for example, a study published by the Water Resources Research Institute, at Louisiana State University, described Old River as "the scene of a direct confrontation between the United States Government and the Mississippi River," and—all constructions of the Corps notwithstanding—awarded the victory to the Mississippi River. "Just when this will occur cannot be predicted," the report concluded. "It could happen next year, during the next decade, or sometime in the next thirty or forty years. But the final outcome is simply a matter of time and it is only prudent to prepare for it."

The Corps thought differently, saying, "We can't let that happen. We are charged by Congress not to let that happen." Its promotional film referred to Old River Control as "a good soldier." Old River Control was, moreover, "the keystone of the comprehensive flood-protection project for the lower Mississippi Valley," and nothing was going to remove the keystone. People arriving at New Orleans District Headquarters, U.S. Army Corps of Engineers, were confronted at the door by a muralled collage of maps and pictures and bold letters unequivocally declaring, "The Old River Control Structures, located about two hundred miles above New Orleans on the Mississippi River, prevent the Mississippi from changing course by controlling flows diverted into the Atchafalaya Basin."

No one's opinions were based on more intimate knowledge than those of LeRoy Dugas, Rabalais's upstream counterpart—the manager of the apparatus that controlled the flow at Old River. Like Rabalais, he was Acadian and of the country. Dugie—as he is universally called—had worked at Old River Control since 1963, when the water started flowing. In years to follow, colonels and generals would seek his counsel. "Those professors at L.S.U. say that whatever we do we're going to lose the system," he remarked one day at Old River, and, after a pause, added, "Maybe they're right." His voice had the sound of water over rock. In pitch, it was lower than a helicon tuba. Better to hear him indoors, in his operations office, away from the structure's competing thunders. "Maybe they're right," he repeated. "We feel that we can hold the river. We're going to try. Whenever you try to control nature, you've got one strike against you."

Dugie's face, weathered and deeply tanned, was saved from looking weary by the alertness and the humor in his eyes. He wore a large, lettered belt buckle that said TO HELP CONTROL THE MISSISSIPPI. "I was originally born in Morganza," he told me. "Thirty miles down the road. I have lived in Pointe Coupee Parish all my life. Once, I even closed my domicile and went to work in Texas for the Corps—but you always come back." (Rabalais also—as he puts it—"left out of here one time," but not for long.) All through Dugie's youth, of course, the Mississippi had spilled out freely to feed the Atchafalaya. He took the vagaries of the waters for granted, not to mention the supremacy of their force in flood. He was a naval gunner on Liberty ships in the South Pacific during the Second World War, and within a year or two of his return was astonished to hear that the Corps of Engineers was planning to restrain Old River. "They were going to try to control the flow," he said. "I thought they had lost their marbles."

Outside, on the roadway that crosses the five-hundred-

and-sixty-six-foot structure, one could readily understand where the marbles might have gone. Even at this time of modest normal flow, we looked down into a rage of water. It was running at about twelve miles an hour—significantly faster than the Yukon after breakup—and it was pounding into the so-called stilling basin on the downstream side, the least still place you would ever see. The No. 10 rapids of the Grand Canyon, which cannot be run without risk of life, resemble the Old River stilling basin, but the rapids of the canyon are a fifth as wide. The Susitna River is sometimes more like it— melted glacier ice from the Alaska Range. Huge trucks full of hardwood logs kept coming from the north to cross the structure, on their way to a chipping mill at Simmesport. One could scarcely hear them as they went by.

There was a high sill next to this one—a separate weir, two-thirds of a mile long and set two feet above the local flood stage, its purpose being to help regulate the flow of extremely high waters. The low sill, as the one we stood on was frequently called, was the prime valve at Old River, and dealt with the water every day. The fate of the project had depended on the low sill, and it was what people meant when, as they often did, they simply said "the structure." The structure and the high sill—like the navigation lock downstream—were fitted into the Mississippi's mainline levee. Beyond the sound of the water, the broad low country around these structures was quiet and truly still. Here and again in the fields, pump jacks bobbed for oil. In the river batture—the silt-swept no man's land between waterline and levee—lone egrets sat in trees, waiting for the next cow.

Dugie remarked that he would soon retire, that he felt old and worn down from fighting the river.

I said to him, "All you need is a good flood."

And he said, "Oh, no. Don't talk like that, man. You talk vulgar."

It was odd to look out toward the main-stem Mississippi, scarcely half a mile away, and see its contents spilling sideways, like cornmeal pouring from a hole in a burlap bag. Dugie said that so much water coming out of the Mississippi created a powerful and deceptive draw, something like a vacuum, that could suck in boats of any size. He had seen some big ones up against the structure. In the mid-sixties, a man alone had come down from Wisconsin in a small double-ended vessel with curling ends and tumblehome—a craft that would not have been unfamiliar to the Algonquians, who named the Mississippi. Dugie called this boat "a pirogue." Whatever it was, the man had paddled it all the way from Wisconsin, intent on reaching New Orleans. When he had nearly conquered the Mississippi, however, he was captured by the At-chafalaya. Old River caught him, pulled him off the Mississippi, and shot him through the structure. "He was in shock, but he lived," Dugie said. "We put him in the hospital in Natchez."

After a moment, I said, "This is an exciting place."

And Dugie said, "You've heard of Murphy—'What can happen will happen'? This is where Murphy lives."

A TOWBOAT coming up the Atchafalaya may be running from Corpus Christi to Vicksburg with a cargo of gasoline, or from Houston to St. Paul with ethylene glycol. Occasionally, Rabalais sees a sailboat, more rarely a canoe. One time, a cottonwood-log dugout with a high Viking bow went past Old River. A ship carrying Leif Eriksson himself, however, would be less likely to arrest the undivided attention of the lockmaster than a certain red-trimmed cream-hulled vessel called Mississippi, bearing Major General Thomas Sands.

Each year, in late summer or early fall, the Mississippi comes down its eponymous river and noses into the lock. This is the Low-Water Inspection Trip, when the General makes a journey from St. Louis and into the Atchafalaya, stopping along the way at river towns, picking up visitors, listening to complaints. In external configuration, the Mississippi is a regular towboat—two hundred and seventeen feet long, fifty feet wide, its horsepower approaching four thousand. The term "towboat" is a misnomer, for the river towboats all push their assembled barges and are therefore designed with broad flat bows. Their unpleasant profiles seem precarious, as if they were the rear halves of ships that have been cut in two. The Mississippi triumphs over these disadvantages. Intended as a carrier of influenceable people, it makes up in luxury what it suffers in form. Only its red trim is martial. Its over-all bright cream suggests globules that have risen to the top. Its broad flat front is a wall of picture windows, of riverine panoramas, faced with cream-colored couches among coffee tables and standing lamps. A river towboat will push as many as fifty barges at one time. What this boat pushes is the program of the Corps.

The Mississippi, on its fall trip, is the site of on-board hearings at Cape Girardeau, Memphis, Vicksburg, and, ultimately, Morgan City. Customarily, it arrives at Old River early in the morning. Before the boat goes through the lock, people with names like Broussard, Brignac, Begnaud, Blanchard, Juneau, Gautreau, Caillouet, and Smith get on—people from the Atchafalaya Basin Levee Board, the East Jefferson Levee Board, the Pontchartrain Levee Board, the Louisiana Office of Public Works, the United States Fish and Wildlife

Service, the Teche-Vermilion Fresh Water District. Oliver Houck, the Tulane professor, gets on, and nine people—seven civilians and two colonels—from the New Orleans District of the Corps of Engineers. "This is the ultimate in communications," says the enthusiastic General Sands as he greets his colleagues and guests. The gates close behind the Mississippi. The mooring bitts inside the lock wail like coyotes as the water and the boat go down.

The pilothouse of the Mississippi is a wide handsome room directly above the lounge and similarly fronted with a wall of windows. It has map-and-chart tables, consoles of electronic equipment, redundant radars. The pilots stand front and center, as trim and trig as pilots of the air—John Dugger, from Collierville, Tennessee (the ship's home port is Memphis), and Jorge Cano, a local "contact pilot," who is here to help the regular pilots sense the shoals of the Atchafalaya. Among the mutating profiles of the river, their work is complicated. Mark Twain wrote of river pilots, "Two things seemed pretty apparent to me. One was, that in order to be a pilot a man had got to learn more than any one man ought to be allowed to know; and the other was, that he must learn it all over again in a different way every twenty-four hours. . . . Your true pilot cares nothing about anything on earth but the river, and his pride in his occupation surpasses the pride of kings." Cano, for his part, is somewhat less flattering on the subject of Twain. He says it baffles him that Twain has "such a big reputation for someone who spent so little time on the river." Today, the Atchafalaya waters are twelve feet lower than the Mississippi's. Cano says that the difference is often as much as twenty. Now the gates slowly open, revealing the outflow channel that leads into old Old River and soon to the Atchafalaya.

The Mississippi River Commission, which is part civilian and part military, with General Sands as president, is required by statute to make these trips—to inspect the flood-control and navigation systems from Illinois to the Gulf, and to hold the hearings. Accordingly, there are two major generals and one brigadier aboard, several colonels, various majors—in all, a military concentration that is actually untypical of the U.S. Army Corps of Engineers. The Corps consists essentially of civilians, with a veneer of military people at and near the top. For example, Sands has with him his chief executive assistant, his chief engineer, his chief planner, his chief of operations, and his chief of programming. All these chiefs are civilians. Sands is commander of the Corps' Lower Mississippi Valley Division, of which the New Orleans District, which includes Old River, is a part. The New Orleans District, U.S. Army Corps of Engineers, consists of something like ten Army officers and fourteen hundred civilians.

Just why the Army should be involved at all with levee systems, navigation locks, rock jetties, concrete revetments, and the austere realities of deltaic geomorphology is a question that attracts no obvious answer. The Corps is here because it is here. Its presence is an expression not of contemporary military strategy but of pure evolutionary tradition, its depth of origin about a century and three-quarters. The Corps is here specifically to safeguard the nation against any repetition of the War of 1812. When that unusual year was in its thirty-sixth month, the British Army landed on the Gulf Coast and marched against New Orleans. The war had been promoted, not to say provoked, by territorially aggressive American Mid-westerners who were known around the country as hawks. It had so far produced some invigorating American moments ("We have met the enemy and they are ours"), including

significant naval victories by ships like the Hornet and the
Wasp. By and large, though, the triumphs had been British.
The British had repelled numerous assaults on Canada. They
had established a base in Maine. In Washington, they had
burned the Capitol and the White House, and with their
rutilant rockets and air-burst ballistics they tried to destroy
Baltimore. New Orleans was not unaware of these events, and
very much dreaded invasion. When it came, militarily un-
trained American backwoods sharpshooters, standing behind
things like cotton bales, picked off two thousand soldiers of
the King while losing seventy-one of their own. Nonetheless,
the city's fear of invasion long outlasted the war.

Despite the Treaty of Ghent, there was a widespread as-
sumption that the British would attack again and, if so, would
surely attack where they had attacked before. One did not have
to go to the War College to learn that lightning enjoys a second
chance. Fortifications were therefore required in the environs
of New Orleans. That this was an assignment for the Army
Corps of Engineers was obvious in more than a military sense.
There was—and for another decade would be—only one
school of engineering in America. This was the United States
Military Academy, at West Point, New York. The academy
had been founded in 1802. The beginnings of the Army Corps
of Engineers actually date to the American Revolution. General Washington, finding among his aroused colonists few
engineers worthy of the word, hired engineers from Louis XVI,
and the first Corps was for the most part French.

The Army engineers chose half a dozen sites near New
Orleans and, setting a pattern, signed up a civilian contractor
to build the fortifications. Congress also instructed the Army
to survey the Mississippi and its tributaries with an eye to
assuring and improving inland navigation. Thus the Corps

spread northward from its military fortifications into civil works along the rivers. In the eighteen-forties and fifties, many of these projects were advanced under the supervision of Pierre Gustave Toutant Beauregard, West Point '38, a native of St. Bernard Parish, and ranking military engineer in the district. Late in 1860, Beauregard was named superintendent of the United States Military Academy. He served five days, resigned to become a Confederate general, and opened the Civil War by directing the bombardment of Fort Sumter.

So much for why there are military officers on the towboat Mississippi inspecting the flood controls of Louisiana's delta plain. Thomas Sands—with his two stars, his warm smile, his intuitive sense of people, and his knowledge of hydrology—is Pierre Gustave Toutant Beauregard's apostolic successor. Sands is trim, athletic, and, in appearance, youthful. Only in his Vietnam ribbons does he show the effects of his assignments as a combat engineer. One of his thumbs is larger and less straight than the other, but that is nothing more than an orthopedic reference to the rigors of plebe lacrosse—West Point '58. He grew up near Nashville, and has an advanced degree in hydrology from Texas A. & M. and a law degree he earned at night while working in the Pentagon. As a colonel, he spent three years in charge of the New Orleans District. As a brigadier general, he was commander of the Corps' North Atlantic Division, covering military and civil works from Maine to Virginia. Now, from his division headquarters, in Vicksburg, he is in charge of the Mississippi Valley from Missouri to the Gulf. On a wall of his private office is a board of green slate. One day when I was interviewing him there, he spent much of the time making and erasing chalk diagrams. "Man against nature. That's what life's all about," he said as he sketched the concatenating forces at Old River and the controls the Corps had applied. He used only the middle third of the slate. The rest had been preempted. The words "BE INNOVATIVE, BE RESPONSIVE, AND OPERATE WITH A TOUCH OF CLASS" were chalked across the bottom. "Old River is a true representation of a confrontation with nature," he went on. "Folks recognized that Mother Nature, being what she is—having changed course many times—would do it again. Today, Mother Nature is working within a constrained environment in the lower Mississippi. Old River is the key element. Every facet of law below there relates to what goes on in this little out-of-the-way point that most folks have never heard about." Chalked across the upper third of the slate were the words "DO WHAT'S RIGHT, AND BE PREPARED TO FIGHT AS INFANTRY WHEN REQUIRED!!!"

Now, aboard the towboat Mississippi, the General is saying, "In terms of hydrology, what we've done here at Old River is stop time. We have, in effect, stopped time in terms of the distribution of flows. Man is directing the maturing process of the Atchafalaya and the lower Mississippi." There is nothing formal about these remarks. The General says that this journey downriver is meant to be "a floating convention." Listening to him is not a requirement. From the pilothouse to the fantail, people wander where they please, stopping here and again to converse in small groups.

Two floatplanes appear above the trees, descend, flare, and land side by side behind the Mississippi. The towboat reduces power, and the airplanes taxi into its wake. They carry four passengers from Morgan City—latecomers to the floating convention. They climb aboard, and the airplanes fly away. These four, making such effort to advance their special interests, are four among two million nine hundred thousand people whose livelihoods, safety, health, and quality of life are directly influenced by the Corps' controls at Old River. In years gone by, when there were no control structures, naturally there were no complaints. The water went where it pleased. People took it as it came. The delta was in a state of nature. But now that Old River is valved and metered there are two million nine hundred thousand potential complainers, very few of whom are reluctant to present a grievance to the Corps. When farmers want less water, for example, fishermen want more, and they all complain to the Corps. In General Sands' words, "We're always walkin' around with, by and large, the black hat on. There's no place in the U.S. where there are so many competing interests relating to one water resource."

Aboard the Mississippi, this is the primary theme. Oliver Houck, professor of ecoprudence, is heard to mutter, "What the Corps does with the water decides everything." And General Sands cheerfully remarks that every time he makes one of these trips he gets "beaten on the head and shoulders." He continues, "In most water-resources stories, you can identify two sides. Here there are many more. The crawfisherman and the shrimper come up within five minutes asking for opposite things. The crawfishermen say, 'Put more water in, the water is low.' Shrimpers don't want more water. They are benefitted by low water. Navigation interests say, 'The water is too low, don't take more away or you'll have to dredge.' Municipal interests say, 'Keep the water high or you'll increase saltwater intrusion.' In the high-water season, everybody is interested in less water. As the water starts dropping, upstream farmers say, 'Get the water off of us quicker.' But folks downstream don't want it quicker. As water levels go up, we divert some fresh water into marshes, because the marshes need it for the nutrients and the sedimentation, but oyster fishermen complain. They all complain except the ones who have seed-oyster beds, which are destroyed by excessive salinity. The variety of competing influences is phenomenal."

In southern Louisiana, the bed of the Mississippi River is so far below sea level that a flow of at least a hundred and twenty thousand cubic feet per second is needed to hold back salt water and keep it below New Orleans, which drinks the river. Along the ragged edges of the Gulf, whole ecosystems depend on the relationship of fresh to salt water, which is in large part controlled by the Corps. Shrimp people want water to be brackish, waterfowl people want it fresh—a situation that causes National Marine Fisheries to do battle with United States Fish and Wildlife while both simultaneously attack the Corps. The industrial interests of the American Ruhr beseech the Corps to maintain their supply of fresh water. Agricultural pumping stations demand more fresh water for their rice but nervily ask the Corps to keep the sediment. Morgan City needs water to get oil boats and barges to rigs offshore, but if Morgan City gets too much water it's the end of Morgan City. Port authorities present special needs, and the owners of grain elevators, and the owners of coal elevators, barge interests, flood-control districts, levee boards. As General Sands says, finishing the list, "A guy who wants to put a new dock in has to come to us." People suspect the Corps of favoring other people. In addition to all the things the Corps actually does and does not do, there are infinite actions it is imagined to do, infinite actions it is imagined not to do, and infinite actions it is imagined to be capable of doing, because the Corps has been conceded the almighty role of God.

The towboat enters the Atchafalaya at an unprepossessing T in a jungle of phreatophytic trees. Atchafalaya. The "a"s are broad, the word rhymes with "jambalaya," and the accents are on the second and fourth syllables. Among navigable rivers, the Atchafalaya is widely described as one of the most treacherous in the world, but it just lies there quiet and smooth. It lies there like a big alligator in a low slough, with time on its side, waiting—waiting to outwait the Corps of Engineers—and hunkering down ever lower in its bed and presenting a sort of maw to the Mississippi, into which the river could fall. In the pilothouse, standing behind Jorge Cano and John Dugger as they swing the ship to port and head south, I find myself remembering an exchange between Cano and Rabalais a couple of days ago, when Cano was speculating about the Atchafalaya's chances of capturing the Mississippi someday despite all efforts to prevent it from doing so. "Mother Nature is patient," he said. "Mother Nature has more time than we do."

Rabalais said, "She has nothing but time."

Frederic Chatry happens to be in the pilothouse, too, as does Fred Bayley. Both are civilians: Chatry, chief engineer of the New Orleans District; Bayley, chief engineer of the Lower Mississippi Valley Division. Chatry is short and slender, a courtly and formal man, his uniform a bow tie. He is saying that before the control structures were built water used to flow in either direction through Old River. It would flow into the Mississippi if the Red happened to be higher. This was known as a reversal, and the last reversal occurred in 1945. The enlarging Atchafalaya was by then so powerful in its draw that it took all of the Red and kept it. "The more water the Atchafalaya takes, the bigger it gets; the bigger it gets, the more water it takes. The only thing that interrupts it is Old River Control. If we had not interrupted it, the main river would now be the Atchafalaya, below this point. If you left it to its own devices, the end result had to be that it would become the master stream. If that were to happen, below Old River the Mississippi reach would be unstable. Silt would fill it in. The Corps could not cope with it. Old River to Baton Rouge would fill in. River traffic from the north would stop. Everything would go to pot in the delta. We couldn't cope. It would be plugged."

I ask to what extent they ever contemplate that the structures at Old River might fail.

Bayley is quick to answer—Fred Bayley, a handsome sandy-haired man in a regimental tie and a cool tan suit, with the contemplative manner of an academic and none of the defenses of a challenged engineer. "Anything can fail," he says. "In most of our projects, we try to train natural effects instead of taking them head on. I never approach anything we do with the idea that it can't fail. That is sticking your head in the sand."

We are making twelve knots on a two-and-a-half-knot current under bright sun and cottony bits of cloud—flying along between the Atchafalaya levees, between the river-batture trees. We are running down the reach above Simmesport, but only a distant bridge attests to that fact. From the river you cannot see the country. From the country you cannot see the river. I once looked down at this country from the air, in a light plane, and although it is called a floodway—this segment of it the West Atchafalaya Floodway—it is full of agriculture, in plowed geometries of brown, green, and tan. The Atchafalaya from above looks like the Connecticut winding past New Hampshire floodplain farms. If you look up, you do not see Mt. Washington. You see artificial ponds, now and again, as far as the horizon—square ponds, dotted with the cages of crawfish. You see dark-green pastureland, rail fences, cows with short fat shadows.

The unexpected happens—unthinkable, unfortunate, but

not unimaginable. At first with a modest lurch, and then with a more pronounced lurch, and then with a profound structural shudder, the Mississippi is captured by the Atchafalaya. The mid-American flagship of the U.S. Army Corps of Engineers has run aground.

AFTER GOING ON LINE, in 1963, the control structures at Old River had to wait ten years to prove what they could do. The nineteen-fifties and nineteen-sixties were secure in the Mississippi Valley. In human terms, a generation passed with no disastrous floods. The Mississippi River and Tributaries Project—the Corps' total repertory of defenses from Cairo, Illinois, southward—seemed to have met its design purpose: to confine and conduct the run of the river, to see it safely into the Gulf. The Corps looked upon this accomplishment with understandable pride and, without intended diminution of respect for its enemy, issued a statement of victory: "We harnessed it, straightened it, regularized it, shackled it."

Then, in the fall of 1972, the winter of 1973, river stages were higher than normal, reducing the system's tolerance for what might come in spring. In the upper valley, snows were unusually heavy. In the South came a season of exceptional rains. During the uneventful era that was about to end, the Mississippi's main channel, in its relative lethargy, had given up a lot of volume to accumulations of sediment. High water, therefore, would flow that much higher. As the spring runoff came down the tributaries, collected, and approached, computers gave warning that the mainline levees were not sufficient to contain it. Eight hundred miles of frantically filled sandbags were added to the levees. Bulldozers added potato ridges— barriers of uncompacted dirt. While this was going on, more rain was falling. In the southern part of the valley, twenty inches fell in a day and a half.

At Old River Control on an ordinary day, when the stilling basin sounds like Victoria Falls but otherwise the country is calm and dry—when sandy spaces and stands of trees fill up the view between the structure and the Mississippi—an almost academic effort is required to visualize a slab of water six stories high, spread to the ends of perspective. That is how it was in 1973. During the sustained spring high water—week after week after week—the gathered drainage of Middle America came to Old River in units exceeding two million cubic feet a second. Twenty-five per cent of that left the Mississippi channel and went to the Atchafalaya. In aerial view, trees and fields were no longer visible, and the gated stronghold of the Corps seemed vulnerable in the extreme—a narrow causeway, a thin fragile line across a brown sea.

The Corps had built Old River Control to control just about as much as was passing through it. In mid-March, when the volume began to approach that amount, curiosity got the best of Raphael G. Kazmann, author of a book called "Modern Hydrology" and professor of civil engineering at Louisiana State University. Kazmann got into his car, crossed the Mississippi on the high bridge at Baton Rouge, and made his way north to Old River. He parked, got out, and began to walk the structure. An extremely low percentage of its five hundred and sixty-six feet eradicated his curiosity. "That whole miserable structure was vibrating," he recalled years later, adding that he had felt as if he were standing on a platform at a small rural train station when "a fully loaded freight goes through." Kazmann opted not to wait for the caboose. "I thought, This thing weighs two hundred thousand tons. When two hundred thousand tons vibrates like this, this is no place for R. G.

Kazmann. I got into my car, turned around, and got the hell out of there. I was just a professor—and, thank God, not responsible."

Kazmann says that the Tennessee River and the Missouri River were "the two main culprits" in the 1973 flood. In one high water and another, the big contributors vary around the watershed. An ultimate deluge might possibly involve them all. After Kazmann went home from Old River that time in 1973, he did his potamology indoors for a while, assembling daily figures. In some of the numbers he felt severe vibrations. In his words, "I watched the Ohio like a hawk, because if that had come up, I thought, Katic, bar the door!"

The water was plenty high as it was, and continuously raged through the structure. Nowhere in the Mississippi Valley were velocities greater than in this one place, where the waters made their hydraulic jump, plunging over what Kazmann describes as "concrete falls" into the regime of the Atchafalaya. The structure and its stilling basin had been configured to dissipate energy—but not nearly so much energy. The excess force was attacking the environment of the structure. A large eddy had formed. Unbeknownst to anyone, its swirling power was excavating sediments by the inflow apron of the structure. Even larger holes had formed under the apron itself. Unfortunately, the main force of the Mississippi was crashing against the south side of the inflow channel, producing unplanned turbulence. The control structure had been set up near the outside of a bend of the river, and closer to the Mississippi than many engineers thought wise.

On the outflow side—where the water fell to the level of the Atchafalaya—a hole had developed that was larger and deeper than a football stadium, and with much the same shape. It was hidden, of course, far beneath the chop of wild water.

The Corps had long since been compelled to leave all eleven gates wide open, in order to reduce to the greatest extent possible the force that was shaking the structure, and so there was no alternative to aggravating the effects on the bed of the channel. In addition to the structure's weight, what was holding it in place was a millipede of stilts—steel H-beams that reached down at various angles, as pilings, ninety feet through sands and silts, through clayey peats and organic mucks. There never was a question of anchoring such a fortress in rock. The shallowest rock was seven thousand feet straight down. In three places below the structure, sheet steel went into the substrate like fins; but the integrity of the structure depended essentially on the H-beams, and vehicular traffic continued to cross it en route to San Luis Rey.

Then, as now, LeRoy Dugas was the person whose hand controlled Old River Control—a thought that makes him smile. "We couldn't afford to close any of the gates," he remarked to me one day at Old River. "Too much water was passing through the structure. Water picked up riprap off the bottom in front, and rammed it through to the tail bed." The riprap included derrick stones, and each stone weighed seven tons. On the level of the road deck, the vibrations increased. The operator of a moving crane let the crane move without him and waited for it at the end of the structure. Dugie continued, "You could get on the structure with your automobile and open the door and it would close the door." The crisis recalled the magnitude of "the '27 high water," when Dugie was a baby. Up the valley somewhere, during the '27 high water, was a railroad bridge with a train sitting on it loaded with coal. The train had been put there because its weight might help keep the bridge in place, but the bridge, vibrating in the floodwater, produced so much friction that the coal in the gondolas caught fire. Soon the bridge, the train, and the glowing coal fell into the water.

One April evening in 1973—at the height of the flood— a fisherman walked onto the structure. There is, after all, order in the universe, and some things take precedence over impending disasters. On the inflow side, facing the Mississippi, the structure was bracketed by a pair of guide walls that reached out like curving arms to bring in the water. Close by the guide wall at the south end was the swirling eddy, which by now had become a whirlpool. There was other motion as well— or so it seemed. The fisherman went to find Dugas, in his command post at the north end of the structure, and told him the guide wall had moved. Dugie told the fisherman he was seeing things. The fisherman nodded affirmatively.

When Dugie himself went to look at the guide wall, he looked at it for the last time. "It was slipping into the river, into the inflow channel." Slowly it dipped, sank, broke. Its foundations were gone. There was nothing below it but water. Professor Kazmann likes to say that this was when the Corps became "scared green." Whatever the engineers may have felt, as soon as the water began to recede they set about learning the dimensions of the damage. The structure was obviously undermined, but how much so, and where? What was solid, what was not? What was directly below the gates and the roadway? With a diamond drill, in a central position, they bored the first of many holes in the structure. When they had penetrated to basal levels, they lowered a television camera into the hole. They saw fish.

THIS WAS SCARCELY the first time that an attempt to control the Mississippi had failed. Old River, 1973, was merely the most emblematic place and moment where, in the course of three centuries, failure had occurred. From the beginnings of settlement, failure was the par expectation with respect to the river—a fact generally masked by the powerful fabric of ambition that impelled people to build towns and cities where almost any camper would be loath to pitch a tent.

If you travel by canoe through the river swamps of Louisiana, you may very well grow uneasy as the sun is going down. You look around for a site—a place to sleep, a place to cook. There is no terra firma. Nothing is solider than duckweed, resting on the water like green burlap. Quietly, you slide through the forest, breaking out now and again into acreages of open lake. You study the dusk for some dark cap of uncovered ground. Seeing one at last, you occupy it, limited though it may be. Your tent site may be smaller than your tent, but in this amphibious milieu you have found yourself terrain. You have established yourself in much the same manner that the French established New Orleans. So what does it matter if your leg spends the night in the water?

The water is from the state of New York, the state of Montana, the province of Alberta, and everywhere below that frame. Far above Old River are places where the floodplain is more than a hundred miles wide. Spaniards in the sixteenth century came upon it at the wrong time, saw an ocean moving south, and may have been discouraged. Where the delta began, at Old River, the water spread out even more—through a palimpsest of bayous and distributary streams in forested paludal basins—but this did not dissuade the French. For military and commercial purposes, they wanted a city in such country. They laid it out in 1718, only months before a great flood. Even as New Orleans was rising, its foundations filled with water. The message in the landscape could not have been more clear: like the aboriginal people, you could fish and forage and move on, but you could not build there—you could not create a city, or even a cluster of modest steadings—without declaring war on nature. You did not have to be Dutch to understand this, or French to ignore it. The people of southern Louisiana have often been compared unfavorably with farmers of the pre-Aswan Nile, who lived on high ground, farmed low ground, and permitted floods to come and go according to the rhythms of nature. There were differences in Louisiana, though. There was no high ground worth mentioning, and planters had to live on their plantations. The waters of the Nile were warm; the Mississippi brought cold northern floods that sometimes stood for months, defeating agriculture for the year. If people were to farm successfully in the rich loams of the natural levees—or anywhere nearby—they could not allow the Mississippi to continue in its natural state. Herbert Kassner, the division's public-relations director, once remarked to me, "This river used to meander all over its floodplain. People would move their tepees, and that was that. You can't move Vicksburg."

When rivers go over their banks, the spreading water immediately slows up, dropping the heavier sediments. The finer the silt, the farther it is scattered, but so much falls close to the river that natural levees rise through time. The first houses of New Orleans were built on the natural levees, overlooking the river. In the face of disaster, there was no better place to go. If there was to be a New Orleans, the levees themselves would have to be raised, and the owners of the houses were ordered to do the raising. This law (1724) was about as effective as the ordinances that compel homeowners and shopkeepers in the North to shovel snow off their sidewalks. Odd as it seems now, those early levees were only three feet high, and they were rife with imperfections. To the extent that they were effective at all, they owed a great deal to the country across the river, where there were no artificial levees, and waters that went over the bank flowed to the horizon. In 1727, the French colonial governor declared the New Orleans levee complete, adding that within a year it would be extended a number of miles up and down the river, making the community floodproof. The governor's name was Perrier. If words could stop water, Perrier had found them—initiating a durable genre.

In 1735, New Orleans went under—and again in 1785. The intervals—like those between earthquakes in San Francisco—were generally long enough to allow the people to build up a false sense of security. In response to the major floods, they extended and raised the levees. A levee appeared across the river from New Orleans, and by 1812 the west bank was leveed to the vicinity of Old River, a couple of hundred miles upstream. At that time, the east bank was leveed as far as Baton Rouge. Neither of the levees was continuous. Both protected plantation land. Where the country remained as the Choctaws had known it, floodwaters poured to the side, reducing the threat elsewhere. Land was not cheap—forty acres cost three thousand dollars—but so great was the demand for riverfront plantations that by 1828 the levees in southern Louisiana were continuous, the river artificially confined. Just in case the levees should fail, some plantation houses—among their fields of sugarcane, their long bright rows of oranges—were built on Indian burial mounds. In 1828, Bayou Manchac was closed. In the whole of the Mississippi's delta plain, Bayou Manchac happened to have been the only distributary that went east. It was dammed at the source. Its discharge would no longer ease the pressures of the master stream.

By this time, Henry Shreve had appeared on the scene—
in various ways to change it forever. He was the consummate
riverman: boatman, pilot, entrepreneur, empirical naval architect. He is noted as the creator of the flat-hulled layer-cake
stern-wheel Mississippi steamboat, its shallow draft the result
of moving the machinery up from below to occupy its own
deck. The Mississippi steamboat was not invented, however.
It evolved. And Shreve's contribution was less in its configuration than its power. A steamboat built and piloted by Henry
Shreve travelled north against the current as far as Louisville.
He demonstrated that commerce could go both ways. Navigation was inconvenienced, though, by hazards in the river—
the worst of which were huge trees that had drifted south over
the years and become stuck in various ways. One kind was
rigid in the riverbed and stood up like a spear. It was called a
planter. Another, known as a sawyer, sawed up and down with
the vagaries of the current, and was likely to rise suddenly in
the path of a boat and destroy it. In the Yukon River, such
logs—eternally bowing—are known as preachers. In the Mississippi, whatever the arrested logs were called individually,
they were all "snags," and after the Army engineers had made
Shreve, a civilian, their Superintendent of Western River Improvements he went around like a dentist yanking snags. The
multihulled snag boats were devices of his invention. In the
Red River, he undertook to disassemble a "raft"—uprooted
trees by the tens of thousands that were stopping navigation
for a hundred and sixty miles. Shreve cleared eighty miles in
one year. Meanwhile, at 31 degrees north latitude (about halfway between Vicksburg and Baton Rouge) he made a bold
move on the Mississippi. In the sinusoidal path of the river,
any meander tended to grow until its loop was so large it would
cut itself off. At 31 degrees north latitude was a west-bending

loop that was eighteen miles around and had so nearly doubled back upon itself that Shreve decided to help it out. He adapted one of his snag boats as a dredge, and after two weeks of digging across the narrow neck he had a good swift current flowing. The Mississippi quickly took over. The width of Shreve's new channel doubled in two days. A few days more and it had become the main channel of the river.

The great loop at 31 degrees north happened to be where the Red-Atchafalaya conjoined the Mississippi, like a pair of parentheses back to back. Steamboats had had difficulty there in the colliding waters. Shreve's purpose in cutting off the loop was to give the boats a smoother shorter way to go, and, as an incidental, to speed up the Mississippi, lowering, however slightly, its crests in flood. One effect of the cutoff was to increase the flow of water out of the Mississippi and into the Atchafalaya, advancing the date of ultimate capture. Where the flow departed from the Mississippi now, it followed an arm of the cutoff meander. This short body of water soon became known as Old River. In less than a fortnight, it had been removed as a segment of the main-stem Mississippi and restyled as a form of surgical drain.

In city and country, riverfront owners became sensitive about the fact that the levees they were obliged to build were protecting not only their properties but also the properties behind them. Levee districts were established—administered by levee boards—to spread the cost. The more the levees confined the river, the more destructive it became when they failed. A place where water broke through was known as a crevasse—a source of terror no less effective than a bursting dam—and the big ones were memorialized, like other great disasters, in a series of proper names: the Macarty Crevasse (1816), the Sauvé Crevasse (1849). Levee inspectors were given power to call out male slaves—aged fifteen to sixty—whose owners lived within seven miles of trouble. With the approach of mid-century, the levees were averaging six feet—twice their original height—and calculations indicated that the flow line would rise. Most levee districts were not populous enough to cover the multiplying costs, so the United States Congress, in 1850, wrote the Swamp and Overflow Land Act. It is possible that no friend of Peter had ever been so generous in handing over his money to Paul. The federal government deeded millions of acres of swampland to states along the river, and the states sold the acreage to pay for the levees. The Swamp Act gave eight and a half million acres of river swamps and marshes to Louisiana alone. Other states, in aggregate, got twenty million more. Since time immemorial, these river swamps had been the natural reservoirs where floodwaters were taken in and held, and gradually released as the flood went down. Where there was timber (including virgin cypress), the swampland was sold for seventy-five cents an acre, twelve and a half cents where there were no trees. The new owners were for the most part absentee. An absentee was a Yankee. The new owners drained much of the swampland, turned it into farmland, and demanded the protection of new and larger levees. At this point, Congress might have asked itself which was the act and which was the swamp.

River stages, in their wide variations, became generally higher through time, as the water was presented with fewer outlets. People began to wonder if the levees could ever be high enough and strong enough to make the river safe. Possibly a system of dams and reservoirs in the tributaries of the upper valley could hold water back and release it in the drier months, and possibly a system of spillways and floodways could be fashioned in the lower valley to distribute water when big floods arrived. Beginning in the eighteen-fifties, these notions were the subject of virulent debate among civilian and military engineers. Four major floods in ten years and thirty-two disastrous crevasses in a single spring were not enough to suggest to the Corps that levees alone might never be equal to the job. The Corps, as things stood, was not yet in charge. District by district, state by state, the levee system was still a patchwork effort. There was no high command in the fight against the water. In one of the Corps' official histories, the situation is expressed in this rather preoccupied sentence: "By 1860, it had become increasingly obvious that a successful war over such an immense battleground could be waged only by a consolidated army under one authority." While the Civil War came and went, the posture of the river did not change. Vicksburg fell but did not move. In the floods of 1862, 1866, and 1867, levees failed. Catastrophes notwithstanding, Bayou Plaquemine—a major distributary of the Mississippi and a natural escape for large percentages of spring high water—was closed in 1868, its junction with the Mississippi sealed by an earthen dam. Even at normal stages, the Mississippi was beginning to stand up like a large vein on the back of a hand. The river of the eighteen-seventies ran higher than it ever had before.

In 1879, Congress at last created the Mississippi River Commission, which included civilians but granted hegemony to the Corps. The president of the commission would always be an Army engineer, and all decisions were subject to veto by the commandant of the Corps. Imperiously, Congress ordered the commission to "prevent destructive floods," and left it to the Corps to say how. The Corps remained committed to the argument that tributary dams and reservoirs and downstream spillways would create more problems than they would solve. "Hold by levees" was the way to do the job.

The national importance of the commission is perhaps illuminated by the fact that one of its first civilian members was Benjamin Harrison. Another was James B. Eads, probably the most brilliant engineer who has ever addressed his attention to the Mississippi River. As a young man, he had walked around on its bottom under a device of his own invention that he called a submarine. As a naval architect in the Civil War, he had designed the first American ironclads. Later, at St. Louis, he had built the first permanent bridge across the main stem of the river south of the Missouri. More recently, in defiance of the cumulative wisdom of nearly everyone in his profession, he had solved a primal question in anadromous navigation: how to get into the river. The mouth was defended by a mud-lump blockade—impenetrable masses of sediment dumped by the river as it reached the still waters of the Gulf. Dredging was hopeless. What would make a channel deep enough for ships? The government wouldn't finance him, so Eads bet his own considerable fortune on an elegant idea: he built parallel jetties in the river's mouth. They pinched the currents. The accelerated water dug out and maintained a navigable channel.

To the Corps' belief that a river confined by levees would similarly look after itself the success of the jetties gave considerable reinforcement. And Eads added words that spoke louder than his actions. "If the profession of an engineer were not based upon exact science," he said, "I might tremble for the result, in view of the immensity of the interests dependent on my success. But every atom that moves onward in the river, from the moment it leaves its home among the crystal springs or mountain snows, throughout the fifteen hundred leagues of its devious pathway, until it is finally lost in the vast waters of the Gulf, is controlled by laws as fixed and certain as those

(38

which direct the majestic march of the heavenly spheres. Every phenomenon and apparent eccentricity of the river—its scouring and depositing action, its caving banks, the formation of the bars at its mouth, the effect of the waves and tides of the sea upon its currents and deposits—is controlled by law as immutable as the Creator, and the engineer need only to be insured that he does not ignore the existence of any of these laws, to feel positively certain of the results he aims at."

When the commission was created, Mark Twain was forty-three. A book he happened to be working on was "Life on the Mississippi." Through a character called Uncle Mumford, he remarked that "four years at West Point, and plenty of books and schooling, will learn a man a good deal, I reckon, but it won't learn him the river." Twain also wrote, "One who knows the Mississippi will promptly aver—not aloud but to himself—that ten thousand River Commissions, with the mines of the world at their back, cannot tame that lawless stream, cannot curb it or confine it, cannot say to it, 'Go here,' or 'Go there,' and make it obey; cannot save a shore which it has sentenced; cannot bar its path with an obstruction which it will not tear down, dance over, and laugh at. But a discreet man will not put these things into spoken words; for the West Point engineers have not their superiors anywhere; they know all that can be known of their abstruse science; and so, since they conceive that they can fetter and handcuff that river and boss him, it is but wisdom for the unscientific man to keep still, lie low, and wait till they do it. Captain Eads, with his jetties, has done a work at the mouth of the Mississippi which seemed clearly impossible; so we do not feel full confidence now to prophesy against like impossibilities. Otherwise one would pipe out and say the Commission might as well bully the comets in their courses and undertake to make them be-

JOHN MCPHEE

have, as try to bully the Mississippi into right and reasonable
conduct."

In 1882 came the most destructive flood of the nineteenth
century. After breaking the levees in two hundred and eighty-
four crevasses, the water spread out as much as seventy miles.
In the fertile lands on the two sides of Old River, plantations
were deeply submerged, and livestock survived in flatboats. A
floating journalist who reported these scenes in the March 29
New Orleans *Times–Democrat* said, "The current running
down the Atchafalaya was very swift, the Mississippi showing
a predilection in that direction, which needs only to be seen
to enforce the opinion of that river's desperate endeavors to
find a short way to the Gulf." The capture of the Mississippi,
in other words, was already obvious enough to be noticed by
a journalist. Seventy-eight years earlier—just after the Louisiana Purchase—the Army officer who went to take possession
of the new country observed the Atchafalaya "completely ob-
structed by logs and other material" and said in his report,
"Were it not for these obstructions, the probability is that the
Mississippi would soon find a much nearer way to the Gulf
than at present, particularly as it manifests a constant incli-
nation to vary its course." The head of the Atchafalaya was
plugged with logs for thirty miles. The raft was so compact
that El Camino Real, the Spanish trail coming in from Texas,
crossed the Atchafalaya near its head, and cattle being driven
toward the Mississippi walked across the logs. The logjam was
Old River Control Structure No. 0. Gradually, it was disas-
sembled, freeing the Atchafalaya to lower its plain. Snag boats
worked on it, and an attempt was made to clear it with fire.
The flood of 1863 apparently broke it open, and at once the
Atchafalaya began to widen and deepen, increasing its draw
on the Mississippi. Shreve's clearing of the Red River had also

increased the flow of the Atchafalaya. The interventional skill of human engineers, which would be called upon in the twentieth century to stop the great shift at Old River, did much in the nineteenth to hurry it up.

For forty-eight years, the Mississippi River Commission and the Corps of Engineers adhered strictly to the "hold by levees" policy—levees, and levees only. It was important that no water be allowed to escape the river, because its full power would be most effective in scouring the bed, deepening the channel, increasing velocity, lowering stages, and preventing destructive floods. This was the hydraulic and hydrological philosophy not only of the U.S. Army Corps of Engineers but also of the great seventeenth-century savant Domenico Guglielmini, whose insights, ultimately, were to prove so ineffective in the valley of the Po. In 1885, one of General Sands' predecessors said, "The commission is distinctly committed to the idea of closing all outlets. . . . It has consistently opposed the fallacy known as the 'Outlet System.' "

Slaves with wheelbarrows started the levees. Immigrants with wheelbarrows replaced the slaves. Mule-drawn scrapers replaced the wheelbarrows, but not until the twentieth century. Fifteen hundred miles of earthen walls—roughly six, then nine, then twelve feet high, and a hundred feet from side to side—were built by men with shovels. They wove huge mats of willow poles and laid them down in cutbanks as revetments. When floods came, they went out to defend their defenses, and, in the words of a Corps publication, the effort was comparable to "the rigors of the battlefield." Nature was not always the only enemy. Anywhere along the river, people were safer if the levee failed across the way. If you lived on the east side, you might not be sad if water flooded west. You were also safer if the levee broke on your own side downstream. Armed

patrols went up and down the levees. They watched for sand
boils—signs of seepage that could open a crevasse from within.
And they watched for private commandos, landing in the dark
with dynamite.

Bayou Lafourche, a major distributary, was dammed in
1904. In something like twenty years, the increased confinement of the river had elevated floodwaters in Memphis by an
average of about eight feet. The Corps remained loyal to the
teachings of Guglielmini, and pronouncements were still
forthcoming that the river was at last under control and destructive floods would not occur again. Declarations of that
sort had been made in the quiet times before the great floods
of 1884, 1890, 1891, 1897, 1898, and 1903, and they would
be made again before 1912, 1913, 1922, and 1927.

The '27 high water tore the valley apart. On both sides
of the river, levees crevassed from Cairo to the Gulf, and in
the same thousand miles the flood destroyed every bridge. It
killed hundreds of people, thousands of animals. Overbank,
it covered twenty-six thousand square miles. It stayed on the
land as much as three months. New Orleans was saved by
blowing up a levee downstream. Yet the total volume of the
1927 high water was nowhere near a record. It was not a
hundred-year flood. It was a form of explosion, achieved by
the confining levees.

THE LEVEES OF THE nineteen-twenties were about six
times as high as their earliest predecessors, but really no more
effective. In a sense, they had been an empirical experiment—
in aggregate, fifteen hundred miles of trial and error. They
could be—and they would be—raised even higher. But in 1927
the results of the experiment at last came clear. The levees

were helping to aggravate the problem they were meant to solve. With walls alone, one could only build an absurdly elevated aqueduct. Resistance times the resistance distance amplified the force of nature. Every phenomenon and apparent eccentricity of the river might be subject to laws as fixed and certain as those which direct the majestic march of the heavenly spheres, but, if so, the laws were inexactly understood. The Corps had attacked Antaeus without quite knowing who he was.

Congress appropriated three hundred million dollars to find out. This was more money in one bill—the hopefully titled Flood Control Act (1928)—than had been spent on Mississippi levees in all of colonial and American history. These were the start-up funds for the Mississippi River and Tributaries Project, the coordinated defenses that would still be incomplete in the nineteen-eighties and would ultimately cost about seven billion dollars. The project would raise levees and build new ones, pave cutbanks, sever loops to align the current, and hold back large volumes of water with substantial dams in tributary streams. Dredges known as dustpans would take up sediment by the millions of tons. Stone dikes would appear in strategic places, forcing the water to go around them, preventing the channel from spreading out. Most significantly, though, the project would acknowledge the superiority of the force with which it was meant to deal. It would give back to the river some measure of the freedom lost as the delta's distributaries one by one were sealed. It would go into the levees in certain places and build gates that could be opened in times of extraordinary flood. The water coming out of such spillways would enter new systems of levees guiding it down floodways to the Gulf. But how many spillways? How many floodways? How many tributary dams? Calculating maximum storms, frequency of storms, maximum snowmelts, sustained saturation of the upper valley, coincident storms in scattered parts of the watershed, the Corps reached for the figure that would float Noah. The round number was three million—that is, three million cubic feet per second coming past Old River. This was twenty-five per cent above the 1927 high. The expanded control system, with its variety of devices, would have to be designed to process that. Various names were given to this blue-moon superflow, this concatenation of recorded moments written in the future unknown. It was called the Design Flood. Alternatively, it was called the Project Flood.

Bonnet Carre was the first spillway—completed in 1931, roughly thirty miles upriver from New Orleans. The water was meant to spill into Lake Pontchartrain and go on into the Gulf, dispersing eight and a half per cent of the Project Flood. Bonnet Carre (locally pronounced "Bonny Carey") would replace dynamite in the defense of New Orleans. When the great crest of 1937 came down the river—setting an all-time record at Natchez—enough of the new improvements were in place to see it through in relative safety, with the final and supreme test presented at Bonnet Carre, where the gates were opened for the first time. At the high point, more than two hundred thousand feet per second were diverted into Lake Pontchartrain, and the flow that went on by New Orleans left the city low and dry.

For the Corps of Engineers, not to mention the people of the southern parishes, the triumph of 1937 brought fresh courage, renewed confidence—a sense once again that the river could be controlled. Major General Harley B. Ferguson, the division commander, became a regional military hero. It was he who had advocated the project's many cutoffs, all made in the decade since 1927, which shortened the river by more

than a hundred miles, reducing the amount of friction working
against the water. The more distance, the more friction. Friction slows the river and raises its level. The mainline levees
were rebuilt, extended, reinforced—and their height was almost doubled, reaching thirty feet. There was now a Great
Wall of China running up each side of the river, with the
difference that while the levees were each about as long as the
Great Wall they were in many places higher and in cross-
section ten times as large. Work continued on the floodways.
There was one in Missouri that let water out of the river and
put it back into the river a few miles downstream. But the
principal conduit of release—without which Bonnet Carre
would be about as useful as a bailing can—was the route of
the Atchafalaya. Since the lower part of it was the largest river
swamp in North America, it was, by nature, ready for the
storage of water. The Corps built guide levees about seventeen
miles apart to shape the discharge toward Atchafalaya Bay,
incidentally establishing a framework for the swamp. In the
northern Atchafalaya, near Old River, they built a three-
chambered system of floodways involving so many intersecting
levees that the country soon resembled a cranberry farm developed on an epic scale. The West Atchafalaya Floodway had
so many people in it, and so many soybeans, that its levees
were to be breached only by explosives in extreme emergency—maybe once in a hundred years. The Morganza Floodway, completed in the nineteen-fifties, contained farmlands
but no permanent buildings. A couple of towns and the odd
refinery were surrounded by levees in the form of rings. But
the plane geometry of the floodways was primarily intended
to take the water from the Mississippi and get it to the swamp.

The flood-control design of 1928 had left Old River
open—the only distributary of the Mississippi to continue in

JOHN MCPHEE

its natural state. The Army was aware of the threat from the
Atchafalaya. Colonel Charles Potter, president of the Mississippi River Commission, told Congress in 1928 that the Mississippi was "just itching to go that way." In the new master
plan, however, nothing resulted from his testimony. The
Corps, in making its flow diagrams, planned that the Atchafalaya would take nearly half the Mississippi during the Design
Flood. It was not in the design that the Atchafalaya take it all.

The Atchafalaya, continuing to grow, had become, by
volume of discharge, one of the three or four largest rivers in
the United States. Compared with the Mississippi, it had a
three-to-one advantage in slope. Around 1950, geologists predicted that by 1975 the shift would be unstoppable. The Mississippi River and Tributaries Project would be in large part
invalidated, the entire levee system of southern Louisiana
would have to be rebuilt, communities like Morgan City in
the Atchafalaya Basin would be a good deal less preserved than
Pompeii, and the new mouth of the Mississippi would be a
hundred and twenty miles from the old. Old River Control
was authorized in 1954.

The levees were raised again. What had been adequate
in 1937 was problematical in the nineteen-fifties. New grades
were set. New dollars were spent to meet the grades. So often
compared with the Great Wall of China, the levees had more
in common with the Maginot Line. Taken together, they were
a retroactive redoubt, more than adequate to wage a bygone
war but below the requirements of the war to come. The levee
grades of the nineteen-fifties would prove inadequate in the
nineteen-seventies. Every shopping center, every drainage improvement, every square foot of new pavement in nearly half
the United States was accelerating runoff toward Louisiana.
Streams were being channelized to drain swamps. Meanders

were cut off to speed up flow. The valley's natural storage capacities were everywhere reduced. As contributing factors grew, the river delivered more flood for less rain. The precipitation that produced the great flood of 1973 was only about twenty per cent above normal. Yet the crest at St. Louis was the highest ever recorded there. The flood proved that control of the Mississippi was as much a hope for the future as control of the Mississippi had ever been. The 1973 high water did not come close to being a Project Flood. It merely came close to wiping out the project.

While the control structure at Old River was shaking, more than a third of the Mississippi was going down the Atchafalaya. If the structure had toppled, the flow would have risen to seventy per cent. It was enough to scare not only a Louisiana State University professor but the division commander himself. At the time, this was Major General Charles Noble. He walked the bridge, looked down into the exploding water, and later wrote these words: "The south training wall on the Mississippi River side of the structure failed very early in the flood, causing violent eddy patterns and extreme turbulence. The toppled training wall monoliths worsened the situation. The integrity of the structure at this point was greatly in doubt. It was frightening to stand above the gate bays and experience the punishing vibrations caused by the violently turbulent, massive flood waters."

If the General had known what was below him, he might have sounded retreat. The Old River Control Structure—this two-hundred-thousand-ton keystone of the comprehensive flood-protection project for the lower Mississippi Valley—was teetering on steel pilings above extensive cavities full of water. The gates of the Morganza Floodway, thirty miles downstream, had never been opened. The soybean farmers of Morganza were begging the Corps not to open them now. The Corps thought it over for a few days while the Old River Control Structure, absorbing shock of the sort that could bring down a skyscraper, continued to shake. Relieving some of the pressure, the Corps opened Morganza.

The damage at Old River was increased but not initiated by the 1973 flood. The invasive scouring of the channel bed and the undermining of the control structure may actually have begun in 1963, as soon as the structure opened. In years that followed, loose barges now and again slammed against the gates, stuck there for months, blocked the flow, enhanced the hydraulic jump, and no doubt contributed to the scouring. Scour holes formed on both sides of the control structure, and expanded steadily. If they had met in 1973, they might have brought the structure down.

After the waters quieted and the concrete had been penetrated by exploratory diamond drills, Old River Control at once became, and has since remained, the civil-works project of highest national priority for the U.S. Army Corps of Engineers. Through the surface of Louisiana 15, the road that traverses the structure, more holes were drilled, with diameters the size of dinner plates, and grout was inserted in the cavities below, like fillings in a row of molars. The grout was cement and bentonite. The drilling and filling went on for months. There was no alternative to leaving gates open and giving up control. Stress on the structure was lowest with the gates open. Turbulence in the channel was commensurately higher. The greater turbulence allowed the water on the Atchafalaya side to dig deeper and increase its advantage over the Mississippi side. As the Corps has reported, "The percentage of Mississippi River flow being diverted through the structure in the absence of control was steadily increasing." That could not be helped.

After three and a half years, control was to some extent restored, but the extent was limited. In the words of the Corps, "The partial foundation undermining which occurred in 1973 inflicted permanent damage to the foundation of the low sill control structure. Emergency foundation repair, in the form of rock riprap and cement grout, was performed to safeguard the structure from a potential total failure. The foundation under approximately fifty per cent of the structure was drastically and irrevocably changed." The structure had been built to function with a maximum difference of thirty-seven feet between the Mississippi and Atchafalaya sides. That maximum now had to be lowered to twenty-two feet—a diminution that brought forth the humor in the phrase "Old River Control." Robert Fairless, a New Orleans District engineer who has long been a part of the Old River story, once told me that "things were touch and go for some months in 1973" and the situation was precarious still. "At a head greater than twenty-two feet, there's danger of losing the whole thing," he said. "If loose barges were to be pulled into the front of the structure where they would block the flow, the head would build up, and there'd be nothing we could do about it."

A sign appeared on one of the three remaining wing walls:

FISHING AND SHAD DIPPING

OFF THIS WING WALL

IS PROHIBITED

A survey boat, Navy-gray and very powerful and much resembling PT-109, began to make runs toward the sill upstream through the roiling brown rapids. Year after year this has continued. The survey boat drives itself to a standstill in the whaleback waves a few yards shy of the structure. Two men in life vests, who stand on the swaying deck in spray that curls like smoke, let go a fifty-pound ball that drops on a cable from a big stainless reel. The ball sinks to the bottom. The crewmen note the depth. They are not looking for mark twain. For example, in 1974 they found three holes so deep that it took a hundred and eighty-five thousand tons of rock to fill them in.

The 1973 flood shook the control structure a whole lot more than it shook the confidence of the Corps. When a legislative committee seemed worried, a Corps general reassured them, saying, "The Corps of Engineers can make the Mississippi River go anywhere the Corps directs it to go." On display in division headquarters in Vicksburg is a large aerial photograph of a school bus moving along a dry road beside a levee while a Galilee on the other side laps at the levee crown. This picture alone is a triumph for the Corps. Herbert Kassner, the public-relations director and a master of his craft, says of the picture, "Of course, I tell people the school bus may have been loaded with workers going to fix a break in the levee, but it looks good." And of course, after 1973, the flow lines were recomputed and the levees had to be raised. When the river would pool against the stratosphere was only a question of time.

The Washington Post, in an editorial in November of 1980, called attention to the Corps' efforts to prevent the great shift at Old River, and concluded with this paragraph:

Who will win as this slow-motion confrontation between humankind and nature goes on? No one really knows. But after watching Mt. St. Helens and listening to the guesses about its performance, if we had to bet, we would bet on the river.

The Corps had already seen that bet, and was about to bump it, too. Even before the muds were dry from the 1973 flood, Corps engineers had begun building a model of Old River at their Waterways Experiment Station, in Vicksburg. The model was to cover an acre and a half. A model of that size was modest for the Corps. Not far away, it had a fifteen-acre model of the Mississippi drainage, where water flowing in from the dendritic tips could get itself together and attack Louisiana. The scale was one human stride to the mile. In the time it took to say "one Mississippi," if fourteen gallons went past Arkansas City that was a Project Flood. Something like eight and a half gallons was "a high-water event." "It's the ultimate sandbox—these guys have made a profession of the sandbox," Tulane's Oliver Houck has said, with concealed admiration. "They've put the whole river in a sandbox." The Old River model not only helped with repairs, it also showed a need for supplementary fortification. Since the first control structure was irreparably damaged, a second one, nearby, with its own inflow channel from the Mississippi, should establish full control at Old River and take pressure off the original structure in times of high stress.

To refine the engineering of the auxiliary structure, several additional models, with movable beds, were built on a distorted scale. Making the vertical scale larger than the horizontal was believed to eliminate surface-tension problems in simulating the turbulence of a real river. The channel beds were covered with crushed coal—which has half the specific gravity of sand—or with walnut shells, which were thought to be better replicas of channel-protecting rock but had an unfortunate tendency to decay, releasing gas bubbles. In one model, the stilling basin below the new structure was filled with driveway-size limestone gravel, each piece meant to represent a derrick stone six feet thick. After enough water had churned through these models to satisfy the designers, ground was broken at Old River, about a third of a mile from the crippled sill, for the Old River Control Auxiliary Structure, the most advanced weapon ever developed to prevent the capture of a river—a handsome gift to the American Ruhr, worth three hundred million dollars. In Vicksburg, Robert Fletcher—a sturdily built, footballish sort of engineer, who had explained to me about the nutshells, the coal, and the gravel—said of the new structure, "I hope it works."

The Old River Control Auxiliary Structure is a rank of seven towers, each buff with a white crown. They are vertical on the upstream side, and they slope toward the Atchafalaya. Therefore, they resemble flying buttresses facing the Mississippi. The towers are separated by six arciform gates, convex to the Mississippi, and hinged in trunnion blocks secured with steel to carom the force of the river into the core of the structure. Lifted by cables, these tainter gates, as they are called, are about as light and graceful as anything could be that has a composite weight of twenty-six hundred tons. Each of them is sixty-two feet wide. They are the strongest the Corps has ever designed and built. A work of engineering such as a Maillart bridge or a bridge by Christian Menn can outdo some other works of art, because it is not only a gift to the imagination but also structural in the matrix of the world. The auxiliary structure at Old River contains too many working components to be classed with such a bridge, but in grandeur and in profile it would not shame a pharaoh.

The original Old River Control project, going on line in 1963, cost eighty-six million dollars. The works of repair and supplement have extended the full cost of the battle to five hundred million. The disproportion in these figures does, of course, reflect inflation, but to a much greater extent it reflects the price of lessons learned. It reflects the fact that no one is stretching words who says that in 1973 the control structure failed. The new one is not only bigger and better and more costly; also, no doubt, there are redundancies in its engineering in memory of '73.

In 1983 came the third-greatest flood of the twentieth century—a narrow but decisive victory for the Corps. The Old River Control Auxiliary Structure was nothing much by then but a foundation that had recently been poured in dry ground. The grout in the old structure kept Old River stuck together. Across the Mississippi, a few miles downstream, the water rose to a threatening level at Louisiana's maximum-security prison. The prison was protected not only by the mainline levee but also by a ring levee of its own. Nonetheless, as things appeared for a while the water was going to pour into the prison. The state would have to move the prisoners, taking them in buses out into the road system, risking Lord knows what. The state went on its knees before the Corps: Do something. The Corps evaluated the situation and decided to bet the rehabilitation of the control structure against the rehabilitation of the prisoners. By letting more water through the control structure, the Corps caused the water at the prison to go down.

Viewed from five or six thousand feet in the air, the structures at Old River inspire less confidence than they do up close. They seem temporary, fragile, vastly outmatched by the natural world—a lesion in the side of the Mississippi butterflied with surgical tape. Under construction nearby is a large hydropower plant that will take advantage of the head between the two rivers and, among other things, light the city of Vidalia. The channel cut to serve it raises to three the number of artificial outlets opened locally in the side of the Mississippi River, making Old River a complex of canals and artificial islands, and giving it the appearance of a marina. The Corps is officially confident that all this will stay in place, and supports its claim with a good deal more than walnuts. The amount of limestone that has been imported from Kentucky is enough to confuse a geologist. As Fred Chatry once said, "The Corps of Engineers is convinced that the Mississippi River can be convinced to remain where it is."

I once asked Fred Smith, a geologist who works for the Corps at New Orleans District Headquarters, if he thought Old River Control would eventually be overwhelmed. He said, "Capture doesn't have to happen at the control structures. It could happen somewhere else. The river is close to it a little to the north. That whole area is suspect. The Mississippi wants to go west. Nineteen-seventy-three was a forty-year flood. The big one lies out there somewhere—when the structures can't release all the floodwaters and the levee is going to have to give way. That is when the river's going to jump its banks and try to break through."

Geologists in general have declared the capture inevitable, but, of course, they would. They know that in 1852 the Yellow River shifted its course away from the Yellow Sea, establishing a new mouth four hundred miles from the old. They know the stories of catastrophic shifts by the Mekong, the Indus, the Po, the Volga, the Tigris and the Euphrates. The Rosetta branch of the Nile was the main stem of the river three thousand years ago.

Raphael Kazmann, the hydrologic engineer, who is now emeritus at Louisiana State, sat me down in his study in Baton Rouge, instructed me to turn on a tape recorder, and, with reference to Old River Control, said, "I have no fight with the Corps of Engineers. I may be a critic, but I'm not mad at anybody. It's a good design. Don't get me wrong. These guys are the best. If it doesn't work for them, nobody can do it."

A tape recorder was not a necessity for gathering the impression that nobody could do it. "More and more energy is being dissipated there," Kazmann said. "Floods are more frequent. There will be a bigger and bigger differential head as time goes on. It almost went out in '73. Sooner or later, it will be undermined or bypassed—give way. I have a lot of respect for Mother . . . for this alluvial river of ours. I don't want to be around here when it happens."

The Corps would say he won't be.

"Nobody knows where the hundred-year flood is," Kazmann continued. "Perspective should be a minimum of a hundred years. This is an extremely complicated river system altered by works of man. A fifty-year prediction is not reliable. The data have lost their pristine character. It's a mixture of hydrologic events and human events. Floods across the century are getting higher, low stages lower. The Corps of Engineers— they're scared as hell. They don't know what's going to happen. This is planned chaos. The more planning they do, the more chaotic it is. Nobody knows exactly where it's going to end."

THE TOWBOAT MISSISSIPPI has hit the point of a sandbar. The depth finder shows thirty-eight feet—indicating that there are five fathoms of water between the bottom of the hull and the bed of the river. The depth finder is on the port side of the ship, however, and the sandbar to starboard, only a few feet down. Thus the towboat has come to its convulsive stop, breaking the stride of two major generals and bringing state officials and levee boards out to the rail. General Sands, the division commander, has a look on his face which suggests that Hopkins has just scored on Army but Army will win the game. There is some running around, some eye-bulging, some breaths drawn shallower even than the sandbar—but not here in the pilothouse. John Dugger, the pilot, and Jorge Cano, the local contact pilot, reveal on their faces not the least touch of dismay, or even surprise, whatever they may feel. They behave as if it were absolutely routine to be aiming downstream in midcurrent at zero knots. In a sense, that is true, for this is not some minor navigational challenge, like shooting rapids in an aircraft carrier. This is the Atchafalaya River.

A poker player might get out of an analogous situation by reaching toward a sleeve. A basketball player would reverse pivot—shielding the ball, whirling the body in a complete circle to leave the defender flat as a sandbar. John Dugger seems to be both. He has cut the engines, and now—looking interested, and nothing else—he lets the current take the stern and swing it wide. The big boat spins, reverse pivots, comes off the bar, and leaves it behind.

Conversations resume—in the lounge, on the outer decks, in the pilothouse—and inevitably many of them touch on the subject of controls at Old River. General Sands is saying, "Between 1950 and 1973, there was intensification of land use in the lower Mississippi—a whole generation grew up thinking you could grow soybeans here and never get wet. Since '73, Mother Nature has been trying to catch up. There have been seven high-water events since 1973. Now the auxiliary structure gives these folks all the assurance they need that Old River can continue to operate."

I ask if anyone agrees that the Atchafalaya could capture the Mississippi near the control structures and not through them.

General Sands replies, "I don't know that I'm personally smart enough to answer that, but I'd say no."

Lieutenant Colonel Ed Willis asks C. J. Nettles, chief of

operations for the New Orleans District, if he thinks the auxiliary structure will do the job.

Nettles says, "The jury is out on that one," and adds that he is not as confident about it as others are.

At Old River a couple of days ago, near the new structure, Nettles and LeRoy Dugas were looking over a scene full of cargo barges, labor barges, crawling bulldozers, hundreds of yards of articulated concrete mattress revetments recently sunk into place, and millions of tons of new limestone riprap. Nettles asked Dugie how long he thought the new armor would last.

Dugie said, "Two high waters."

General Sands advances a question: "Had man not settled in southern Louisiana, what would it be like today? Under nature's scenario, what would it be like?" And, not waiting for an answer, he supplies one himself: "If only nature were here, people—except for some hunters and fishermen—couldn't exist here."

Under nature's scenario, with many distributaries spreading the floodwaters left and right across the big deltaic plain, virtually the whole region would be covered—with fresh sediments as well as water. In an average year, some two hundred million tons of sediment are in transport in the river. This is where the foreland Rockies go, the western Appalachians. Southern Louisiana is a very large lump of mountain butter, eight miles thick where it rests upon the continental shelf, half that under New Orleans, a mile and a third at Old River. It is the nature of unconsolidated sediments to compact, condense, and crustally sink. So the whole deltaic plain, a superhimalaya upside down, is to varying extents subsiding, as it has been for thousands of years. Until about 1900, the river and its distributaries were able to compensate for the subsidence with the amounts of fresh sediment they spread in flood. Across the centuries, distribution was uneven, as channels shifted and land would sink in one place and fill in somewhere else, but over all the land-building process was net positive. It was abetted by decaying vegetation, which went into the flooded silts and made soil. Vegetation cannot decay unless it grows first, and it grew in large part on nutrients supplied by floodwaters.

"In the seventeenth century, the Mississippi was very porous along its banks, and water left it in many places," Fred Chatry reminds us. "Only at low water was it completely confined. Now, in two thousand miles, the first place where water naturally escapes the Mississippi is at Bayou Baptiste Collette—sixty miles below New Orleans."

What was a net gain before 1900 has by now been a net loss for nearly a hundred years, and the Louisiana we have known—from Old River and the Acadian world to Bayou Baptiste Collette—is sinking. Sediments are being kept within the mainline levees and shot into the Gulf at the rate of three hundred and fifty-six thousand tons a day—shot over the shelf like peas through a peashooter, and lost to the abyssal plain. As waters rise ever higher between levees, the ground behind the levees subsides, with the result that the Mississippi delta plain has become an exaggerated Venice, two hundred miles wide—its rivers, its bayous, its artificial canals a trelliswork of water among subsiding lands.

The medians of interstates are water. St. Bernard Parish, which includes suburbs of New Orleans and is larger than the state of Delaware, is two per cent terra firma, eighteen per cent wetland, and eighty per cent water. A ring levee may surround a whole parish. A ring levee may surround fifty-five square miles of soybeans. Every square foot within a ring levee forces water upward somewhere else.

An Alexander Calder might revel in these motions—interdependent, interconnected, related to the flow at Old River. Calder would have understood Old River Control: the place where the work is attached to the ceiling, and below which everything —New Orleans, Morgan City, the river swamp of the Atchafalaya—dangles and swings.

Something like half of New Orleans is now below sea level—as much as fifteen feet. New Orleans, surrounded by levees, is emplaced between Lake Pontchartrain and the Mississippi like a broad shallow bowl. Nowhere is New Orleans higher than the river's natural bank. Underprivileged people live in the lower elevations, and always have. The rich—by the river—occupy the highest ground. In New Orleans, income and elevation can be correlated on a literally sliding scale: the Garden District on the highest level, Stanley Kowalski in the swamp. The Garden District and its environs are locally known as uptown.

Torrential rains fall on New Orleans—enough to cause flash floods inside the municipal walls. The water has nowhere to go. Left on its own, it would form a lake, rising inexorably from one level of the economy to the next. So it has to be pumped out. Every drop of rain that falls on New Orleans evaporates or is pumped out. Its removal lowers the water table and accelerates the city's subsidence. Where marshes have been drained to create tracts for new housing, ground will shrink, too. People buy landfill to keep up with the Joneses. In the words of Bob Fairless, of the New Orleans District engineers, "It's almost an annual spring ritual to get a load of dirt and fill in the low spots on your lawn." A child jumping up and down on such a lawn can cause the earth to move under another child, on the far side of the lawn.

Many houses are built on slabs that firmly rest on pilings. As the turf around a house gradually subsides, the slab seems to rise. Where the driveway was once flush with the floor of the carport, a bump appears. The front walk sags like a hammock. The sidewalk sags. The bump up to the carport, growing, becomes high enough to knock the front wheels out of alignment. Sakrete appears, like putty beside a windowpane, to ease the bump. The property sinks another foot. The house stays where it is, on its slab and pilings. A ramp is built to get the car into the carport. The ramp rises three feet. But the yard, before long, has subsided four. The carport becomes a porch, with hanging plants and steep wooden steps. A carport that is not firmly anchored may dangle from the side of a house like a third of a drop-leaf table. Under the house, daylight appears. You can see under the slab and out the other side. More landfill or more concrete is packed around the edges to hide the ugly scene. A gas main, broken by the settling earth, leaks below the slab. The sealed cavity fills with gas. The house blows sky high.

"The people cannot have wells, and so they take rainwater," Mark Twain observed in the eighteen-eighties. "Neither can they conveniently have cellars or graves, the town being built upon 'made' ground; so they do without both, and few of the living complain, and none of the others." The others may not complain, but they sometimes leave. New Orleans is not a place for interment. In all its major cemeteries, the clients lie aboveground. In the intramural flash floods, coffins go out of their crypts and take off down the street.

The water in New Orleans' natural aquifer is modest in amount and even less appealing than the water in the river.

The city consumes the effluent of nearly half of America, and, more immediately, of the American Ruhr. None of these matters withstanding, in 1984 New Orleans took first place in the annual Drinking Water Taste Test Challenge of the American Water Works Association.

The river goes through New Orleans like an elevated highway. Jackson Square, in the French Quarter, is on high ground with respect to the rest of New Orleans, but even from the benches of Jackson Square one looks up across the levee at the hulls of passing ships. Their keels are higher than the Astro Turf in the Superdome, and if somehow the ships could turn and move at river level into the city and into the stadium they would hover above the playing field like blimps.

In the early nineteen-eighties, the U.S. Army Corps of Engineers built a new large district headquarters in New Orleans. It is a tetragon, several stories high, with expanses of sheet glass, and it is right beside the river. Its foundation was dug in the mainline levee. That, to a fare-thee-well, is putting your money where your mouth is.

Among the five hundred miles of levee deficiencies now calling for attention along the Mississippi River, the most serious happen to be in New Orleans. Among other factors, the freeboard—the amount of levee that reaches above flood levels—has to be higher in New Orleans to combat the waves of ships. Elsewhere, the deficiencies are averaging between one and two feet with respect to the computed high-water flow line, which goes on rising as runoffs continue to speed up and waters are increasingly confined. Not only is the water higher. The levees tend to sink as well. They press down on the mucks beneath them and squirt materials out to the sides. Their crowns have to be built up. "You put five feet on and three feet sink," a Corps engineer remarked to me one day. This is especially true of the levees that frame the Atchafalaya swamp, so the Corps has given up trying to fight the subsidence there with earth movers alone, and has built concrete floodwalls along the tops of the levees, causing the largest river swamp in North America to appear to be the world's largest prison. It keeps in not only water, of course, but silt. Gradually, the swamp elevations are building up. The people of Acadiana say that the swamp would be the safest place in which to seek refuge in a major flood, because the swamp is higher than the land outside the levees.

As sediments slide down the continental slope and the river is prevented from building a proper lobe—as the delta plain subsides and is not replenished—erosion eats into the coastal marshes, and quantities of Louisiana steadily disappear. The net loss is over fifty square miles a year. In the middle of the nineteenth century, a fort was built about a thousand feet from a saltwater bay east of New Orleans. The fort is now collapsing into the bay. In a hundred years, Louisiana as a whole has decreased by a million acres. Plaquemines Parish is coming to pieces like old rotted cloth. A hundred years hence, there will in all likelihood be no Plaquemines Parish, no Terrebonne Parish. Such losses are being accelerated by access canals to the sites of oil and gas wells. After the canals are dredged, their width increases on its own, and they erode the region from the inside. A typical three-hundred-foot oil-and-gas canal will be six hundred feet wide in five years. There are in Louisiana ten thousand miles of canals. In the nineteen-fifties, after Louisiana had been made nervous by the St. Lawrence Seaway, the Corps of Engineers built the Mississippi River–Gulf Outlet, a shipping canal that saves forty miles by traversing marsh country straight from New Orleans to the Gulf. The canal is known as Mr. Go, and shipping has largely ignored it. Mr. Go, having eroded laterally for twenty-five years, is as much as three times its original width. It has devastated twenty-four thousand acres of wetlands, replacing them with open water. A mile of marsh will reduce a coastal-storm-surge wave by about one inch. Where fifty miles of marsh are gone, fifty inches of additional water will inevitably surge. The Corps has been obliged to deal with this fact by completing the ring of levees around New Orleans, thus creating New Avignon, a walled medieval city accessed by an interstate that jumps over the walls.

"The coast is sinking out of sight," Oliver Houck has said. "We've reversed Mother Nature." Hurricanes greatly advance the coastal erosion process, tearing up landscape made weak by the confinement of the river. The threat of destruction from the south is even greater than the threat from the north.

I went to see Sherwood Gagliano one day—an independent coastal geologist and regional planner who lives in Baton Rouge. "We must recognize that natural processes cannot be restored," he told me. "We can't put it back the way it was. The best we can do is try to get it back in balance, try to treat early symptoms. It's like treating cancer. You get in early, you may do something." Gagliano has urged that water be diverted to compensate for the nutrient starvation and sediment deprivation caused by the levees. In other words, open holes in the riverbank and allow water and sediment to build small deltas into disappearing parishes. "If we don't do these things, we're going to end up with a skeletal framework with levees around it—a set of peninsulas to the Gulf," he said. "We will lose virtually all of our wetlands. The cost of maintaining protected areas will be very high. There will be no buffer between them and the coast."

Professor Kazmann, of L.S.U., seemed less hopeful. He

said, "Attempts to save the coast are pretty much spitting in the ocean."

The Corps is not about to give up the battle, or so much as imagine impending defeat. "Deltas wax and wane," remarks Fred Chatry, in the pilothouse of the Mississippi. "You have to be continuously adjusting the system in consonance with changes that occur." Southern Louisiana may be a house of cards, but, as General Sands suggested, virtually no one would be living in it were it not for the Corps. There is no going back, as Gagliano says—not without going away. And there will be no retreat without a struggle. The Army engineers did not pick this fight. When it started, they were still in France. The guide levees, ring levees, spillways, and floodways that dangle and swing from Old River are here because people, against odds, willed them to be here. Or, as the historian Albert Cowdrey expresses it in the introduction to "Land's End," the Corps' official narrative of its efforts in southern Louisiana, "Society required artifice to survive in a region where nature might reasonably have asked a few more eons to finish a work of creation that was incomplete."

THE TOWBOAT MISSISSIPPI is more than halfway down the Atchafalaya now—beyond the leveed farmland of the upper basin and into the storied swamp. The willows on the two sides of the river, however, continue to be so dense that they block from sight what lies behind them, and all we can see is the unobstructed waterway running on and on, half a mile wide, in filtered sunlight and the shadows of clouds. A southwest breeze has put waves on the water. Broad on the starboard bow, it more than quells the humidity and the heat. Nevertheless, as one might expect, most of the people remain indoors, in the chilled atmosphere of the pilothouse, the coat-and-tie comfort of the lounge. A deck of cards appears, and a game of bouré develops, in showboat motif, among various civilian millionaires—Ed Kyle, of the Morgan City Harbor & Terminal District, dealing off the top to the Pontchartrain Levee Board, the Lafourche Basin Levee Board, the Teche-Vermilion Fresh Water District. Oliver Houck—the law professor, former general counsel of the National Wildlife Federation, whose lone presence signals the continuing existence of the environmental movement—naturally stays outdoors. He has established an eyrie on an upper deck, to windward. Tall and loosely structured, Houck could be a middle-aged high jumper, still in shape to clear six feet. His face in repose is melancholy—made so, perhaps, by the world as his mind would have it in comparison with the world as he sees it. What he is seeing at the moment—in the center of the greatest river swamp in North America, which he and his battalions worked fifteen years to "save"—is a walled-off monotony of sky and water.

General Sands joins him, and they talk easily and informally, as two people will who have faced each other across great quantities of time and paper. Sands remarks again that on inspection trips such as this one he has become used to being "beaten on the head and shoulders" by almost everyone he encounters, not just the odd ecologue attired in alienation.

Houck addresses himself to the head, the shoulders, and the chest, saying that he has deep reservations about Sands' uniform: all those brass trinkets and serried stars, the castle keeps, the stratified ribbons. He says that Sands' habiliments constitute a form of intimidation, especially in a region of the country that has not lost its respect for the military presence. Sands' habiliments are not appropriate in a civilian milieu.

"You are Army—an untypical American entity to be performing a political role like this," Houck says to him, beating on. He tells Sands that he reminds him of "a politician on the stump, going around stroking his constituency." He calls him "a political water czar."

Sands implicitly reminds Houck that if it were not for the U.S. Army Corps of Engineers there wouldn't be any stump, the constituency would be somewhere else, and Houck's neighborhood would be nine feet under water. He says, "Under nature's scenario, think what it would be like."

The water czar, I feel a duty to insert, is not the very model of a major general. If he were to chew nails, he would break his teeth. I am not attempting to suggest that he lacks the presence of a general, or the mien, or the bearing. Yet he is, withal, somewhat less martial than most English teachers. Effusive and friendly in a folk-and-country way, courteous, accommodating, he is of the sort whose upward mobility would be swift in a service industry. Make no mistake, he is a general. "Shall we just go to the Four Seasons? A nice little place to have lunch," he said one day in Vicksburg, and we drove to a large building in the center of town, where his car was left directly in front of the main entrance, beside a bright-yellow curb under various belligerent signs forbidding parking. It stayed there for an hour while he had his crab gumbo.

We approach, on the right, a gap in the Atchafalaya's bank, where the willows open to reveal a plexus of bayous. Houck has been complaining that the old Cajun swamp life of the Atchafalaya Basin is gone now, and has been for many years, as a result of the volumes of water concentrated in the floodway and of rules forbidding people to live inside the levees. "This single piece of plumbing," he says of the Atchafalaya, "is the last great river-overflow swamp in the world and

Atchafalaya

also the biggest floodway in the world—all to protect Baton Rouge and New Orleans." We now come abreast of the gap on the right, and it ends the tedium of the reach upriver. It is a broad window into stands of cypress, their wide fluted bases attached to their reflections in still, dark water. "How I love them," says Houck, who is a conservationist of the sunset school, with legal skills adjunct to the force of his emotion. Pointing into the beauty of the bayous, he informs General Sands, "That's what it's all about."

The General takes in the scene without comment. In silence, we look at the water-standing trees and into narrow passages that disappear among them. They draw me into thoughts of my own. I first went in there in 1980—that is, into the Atchafalaya swamp, away from its floodway levees, and miles from the river. There were four of us, in canoes. The guide was Charles Fryling, a professor of landscape architecture at Louisiana State University, who, among the environmentalists of the eighteenth state, plays Romulus to Oliver Houck's Remus. Fryling is a tall man with a broad forehead, whose hair falls straight to his eyes without the slightest suggestion that comb or brush has ever been invited to intrude upon nature. In 1973, when he moved into his house, on the periphery of Baton Rouge, it sat on a smooth green lawn, in a neighborhood of ranch contemporaries, each on a smooth green lawn. Fryling's yard is now a rough green forest, its sweet gums, grapevines, pepper vines, rattan vines, hackberries, passionflowers, and climbing ferns a showcase of natural succession. In Fryling's words, "It beats the hell out of mowing the lawn." The trees are thirty feet high.

Fryling speaks in a slow country roll that could win him a job in movies. He would be Li'l Abner, or Candide at Fort Dix—the soldier who appears slow in basic training and dies on an intelligence mission twenty-five miles behind enemy lines. He is a graduate of the illustrious forestry school of the State University of New York (Syracuse), his advanced degree is from Harvard, and—to continue the escalation—he knows how to get from here to there in the swamp. This is a remarkable feat in seven hundred thousand acres that change so much and so often that they are largely unmappable. Fryling understands the minor bayous. Sometimes they run one way, sometimes the other. The water contains sediment or is clear. "See. The water is clearer. It's coming toward us. It's coming down from Bayou Pigeon. We'll get through."

If you ask him what something is, he knows. It's green hawthorn. It's deciduous holly. It's water privet. It's water elm. It's a water moccasin—there on the branch of that water oak. The moccasin doesn't move. A moccasin never backs off. Dragonflies land on the gunwales. In the Atchafalaya, dragonflies are known as snake doctors. Leaving the open bayou, the canoes turn into the forest and slide among the trunks of cypress under feathery arrowhead crowns. "Young cypress need a couple of years on dry land to get started, but we send so much water through the Atchafalaya that young trees can't get going. So existing cypress are not—as trees are generally thought to be—a renewable resource. We have to protect them in order to have them."

To be in the Atchafalaya is to float among trees under silently flying blue herons, to see the pileated woodpecker, to hope to see an ivorybill, to hear the prothonotary warbler. The barred owl has a speaking voice as guttural as a dog's. It seems to be growling, "Who cooks for you? Who cooks for y'all?" The barred owl—staring from a branch straight down into the canoes—appears to be a parrot in camouflage. In the language of the Longtown Choctaw, "Hacha Falaia" meant "Long

River." (The words are reversed in translation.) Since my first travels with Fryling, those rippling syllables have symbolized for me the bilateral extensions of the phrase "control of nature." Atchafalaya. The word will now come to mind more or less in echo of any struggle against natural forces—heroic or venal, rash or well advised—when human beings conscript themselves to fight against the earth, to take what is not given, to rout the destroying enemy, to surround the base of Mt. Olympus demanding and expecting the surrender of the gods. The Atchafalaya—this most apparently natural of natural worlds, this swamp of the anhinga, swamp of the nocturnal bear—lies between walls, like a zoo. It is utterly dependent on the U.S. Army Corps of Engineers, whose decisions at Old River can cut it dry or fill it with water and silt. Fryling gave me a green-and-white sticker that said "ATCHAFALAYA." I put it in a window of my car. It has been there for many years, causing drivers on the New Jersey Turnpike to veer in close and crowd my lane while staring at a word that signifies collision.

In the Atchafalaya more recently, we came upon a sport fisherman in a skiff called Mon Ark. "There's all kind of land out there now," he said. He meant not only that the wet parts were low but also that the dry parts were growing. In the Atchafalaya, the land comes and goes, but it comes more than it goes. As the overflow swamp of the only remaining distributary in the delta—the only place other than the mouth of the Mississippi where silt can go—the Atchafalaya is silting in. From a light plane at five hundred feet, this is particularly evident as the reflection of the sun races through trees and shoots forth light from the water. The reflection disappears when it crosses the accumulating land. If land accretes from the shore of a lake or a bayou, the new ground belongs to the shore's owner. If it accretes as an island, it belongs to the state—a situation of which Gilbert would be sure to inform Sullivan. Some fifty thousand acres are caught in this tug-of-war. Wet and dry, three-quarters of the Atchafalaya swampland is privately owned. Nearly all the owners are interested less in the swamp than in what may lie beneath it. The conservationists, the Corps, landowners, and recreational interests have worked out a compromise by which all parties putatively get what they want: floodway, fishway, oil field, Eden. From five hundred feet up, the world below is green swamp everywhere, far as the eye can see. The fact is, though, that the eye can't see very far. The biggest river swamp in North America, between its demarcating levees, is seventeen miles wide and sixty miles long. It is about half of what it was when it began at the Mississippi River and went all the way to Bayou Teche.

The old life of the basin is not entirely gone. It is true that people don't collect moss anymore to use in stuffing furniture, true that the great virgin cypresses are away. Their flared stumps remain, like cabins standing in the water. From the beginning of the nineteenth century, Cajuns made their lives and livings in the swamp. Their grocery stores were afloat, and moved among them, camp to camp. It is true all that has vanished, and the Cajuns live outside the levees, but they and others—operating for the most part alone or in pairs—go into the swamp and take twenty-five million dollars' worth of protein out of the water in any given year. The fish alone can average a thousand pounds an acre, and that, according to Fryling, is "more fish than in any other natural water system in the United States"—two and a half times as productive as the Everglades. The fish are not in the conversation, however, when compared with the crawfish.

I know a crawfisherman named Mike Bourque, who lives

in Catahoula. I remember as if it were today running his lines with him. "Watch your hands. Don't put 'em on the side of the boat. 'Cause smash 'em," he said as we went out of Bayou Gravenburg and headed into the trees. His boat was not a canoe, and the object on the stern was no paddle. It was a fifty-horse Mariner, enough for lift-off if the boat had wings. Bourque's brother-in-law was with us. In French, Bourque told him that he was affecting the balance and to shift his position in the boat. Then, addressing me in English, he said, "Watch yourself, I got to jump that log." Ahead of us, half hidden in water hyacinths, was an impressive floating log, with a solid diameter of about two feet. The boat smashed against it, thrust up and over it, with a piercing aluminum screech. The boat was about seventeen feet long. The brother-in-law, Dave Soileau, called it a bateau. Bourque called it a skiff. "French and English—we mix it up," he said. Ordinarily, he works alone, and talks a good deal to himself. "When I talk to myself, I talk in French. When I meet other fishermen, ninety per cent of the time we speak French." If he doesn't know them, he knows where they live, because each town has its accent.

Like everyone else, he calls the hyacinths lilies—water lilies. This densely growing plant—a waterborne kudzu, an exotic from the Orient—has come to plague Southern waterways and spread over marshes like nuclear winter closing many forms of life. That is not the case, however, in the Atchafalaya, where the lilies are good for the crawfish. The young feed on stuff that clings to the roots. On heavy stems, the water hyacinths grow three to four feet high, so a lot of power is needed to get through them. "You'll never see a fisherman with less than a fifty-horse motor."

Bourque moved the skiff from tree to tree as if he were

on snowshoes in a sugarbush emptying buckets of sap. The crawfish cages were chicken-wire pillows with openings at one end. Bourque pulled them out of the water on cords that were tied to the trees, and poured the crawfish into a device that looked something like a roasting pan and was hinged to the side of the boat. He calls it the trough. Open at the inner end, it forms a kind of ramp down which the crawfish crawl until they drop into a bucket. Dead bait fish, dead crawfish, and other detritus remain in the trough, and thus the living creatures winnow themselves from what is thrown away. Snakes are thrown away. Some of the used bait fish have less remaining flesh than skeletons lifted by waiters who work in white gloves. The larger crawfish weigh a quarter of a pound and are nine inches long, with claw spans greater than that. When the bucket is full, the crawfish in their motions seem to simmer at the top. "C'est bon. C'est bon. Où est le sac?" said Bourque, and Soileau handed him a plastic-burlap sack. Containing forty pounds each, the sacks began to pile up. The crawfish lay quiet. When a sack was moved, or even touched, though, the commotion inside sounded like heavy rain.

The boat climbed another log. The engine cavitated. We broke through brush like an elephant. Bourque had been following what he called the driftwood line, where a small change in depth had caused driftwood to linger. To him the swamp topography was as distinctive and varied as the neighborhoods of a city would be to someone else—these subworlds of the Atchafalaya, out past Bayou Gravenburg, on toward the Red Eye Swamp. "This line used to go in back there, but I moved them out in front," he said in a place that seemed much too redundant to have a back or a front. Colored ribbons, which he called flags, helped to distinguish the fishermen's trees, but he could run his lines without them, covering his four hundred cages. He did about sixty an hour. Soileau, using a grain scoop, shovelled dead alewives and compressed pellets of Acadiana Choice Crawfish Bait into each emptied cage, and Bourque returned it to the water. Bourque told Soileau, who is a biologist with the United States Fish and Wildlife Service, to quit the government and come work for him. Soileau said, "For ten dollars a day?"

Bourque said, "Good future. No benefits."

We were in a coulee, which is like a slough but deeper and with slushier muds at the bottom. A cage came up with seventy crawfish, all dead. The cage had been too low in the muck, where the creatures died in an anoxic slurry. They stirred it up themselves. The cage should just lightly touch the bottom, with the closed end slightly raised.

Bourque next pulled up an empty cage. "Somebody helped me out," he remarked, and added that he had occasionally met a thief in the act of raiding one of his cages.

Soileau said, "There's only one thing to do. Go straight to him, board his vessel, and start slugging. There have been no deaths."

Theft was rising in direct proportion to unemployment. Oil companies owned that part of the swamp. Fishermen have, in fact, been arrested for trespass. Fryling's wife, Doris Falkenheiner, defends them in court. Meanwhile, so many fishermen work the watery forest that there is a plastic ribbon on almost every tree. The fishermen say they have to bring their own trees.

We hit another log. We ran between a cypress and its knees. "We're getting up on the ridge," Bourque said, referring to a subtle, invisible feature of the bottom of the swamp. Out of a cage came a white crawfish, a male. (The male has longer arms.) Crawfish are red, white, or blue. The white ones like the sand of the ridge. Blue ones are rare. Bourque sees fewer than twenty a year. Now he was reaching down into the water for a cage that had been separated from its string by another fisherman's motor.

"*Touchez la?*" asked Soileau.

Bourque answered, "Yes." Then he said, "*Ah, bon,*" as he retrieved the cage.

"Are y'all hungry?" Bourque asked.

"I live hungry," said Soileau.

Bourque turned off the motor and we stopped for lunch: ham sandwiches, Royal Crown, Mr. Porker fried cured pork skins. It was seven-thirty in the morning.

We got up around three-thirty and were driving down the levee by four o'clock—in Bourque's pickup, with the skiff behind. Soileau made the comment that the levees were like cancer, because they had to keep growing while they sank into the swamp. After twenty-five miles, we went down a ramp to a boat landing, where forty-one pickups had arrived before us. Roughly five thousand people take crawfish from the swamp, annually trapping twenty-three million pounds.

Now, at lunchtime, as the early-morning sun began to penetrate the trees, we were looking out on one lovely scene, with tupelo and cypress rising from the water, and pollen on the water like pale-green silk. "The best months are Epp Rill and May," Bourque said. "The water might rise in October sometimes. I'll come and try." He was wearing mirrored sunglasses, a soft cap with a buttoned visor, white rubber boots, and yellow rubber overalls slashed at the crotch. Of middle height, blond and fine-featured, he had sandy hair around his ears and a large curl in back, like a breaking wave. His low-sill mustache looked French. He went to St. Martinville High School, as did Soileau, who married the youngest of Bourque's six sisters. In large script below the windows of a drugstore in St. Martinville, a sign says, "*Sidney Dupois Pharmacien—Au Service de la Santé de Votre Famille.*" The Teche *News*, published down the street, has a regular column headlined "PENSE DONC!!" and contains marriage and death notices about people with names like Boudreau, Tesreau, Landreaux, Passeau, Bordagaray, Lajoie, Fournier, Angelle, and Guidry. Bourque was the youngest in his family and the only sibling male. He explains that Cajuns keep going until they get a male, and this was where the Bourques stopped.

Soileau passed the pork skins. Bourque chewed them crunchily. "Crawfish are *écrevisses* in French," he said. "We call them crawfish."

I mentioned that *écrevisses* are cherished by chefs in France.

Soileau said, "I hear you get only three or four."

Bourque had a recipe of which the nouveaux cuisiniers may not have heard. "Sauté onions in butter, then put in fat out of the head for ten or fifteen minutes, then put meat in for a few minutes more," he said. "Salt. Cayenne pepper. Onion tops. What makes the étouffée is the fat. Some people put a little roux in there. You can stretch it like that." Crawfish étouffée: the Cajun quenelle de brochet. The meat is ground, but not to the end of texture. On Easter Sunday morning in Catahoula, the Bourques have a crawfish ball. At least, I thought that's what they were saying until I saw what they did. They boiled a hundred pounds of crawfish. They ate a crimson mountain of condensed lobsters.

Now we were running in Bayou Eugene, which Soileau and Bourque lyrically pronounced in three syllables—"by yooz yen." We came upon a beaver on a floating log. This was not the animal that founded a nation, the alert and agile slapper of the boreal lakes. This was a Louisiana beaver—huge, half asleep, prone like a walrus, a mound of cinnamon fur with nothing much to do but eat. There was no need to dam a thing here. The Corps of Engineers would see to that. The beaver topples trees just to eat the bark. There is no mandate to practice conservation when you are what is being conserved. "A willow branch eaten by a beaver is just as smooth as if it had been sanded," Soileau remarked. "There's nothing prettier than a willow branch eaten by a beaver." Nutria live in the swamp as well. Bourque said that he sees only four or five alligators a year. A friend of his lost a finger to a cottonmouth. "He was walking through thick lilies, very high lilies, to make a road for his pirogue. The snake bit his finger through a glove." Among the crowns of the cypress, a heron flapped by. Bourque called it a *gros bec*. Soileau called it a yellow-crowned night heron. Bourque said, "The *gros bec* is here for the same purpose we are: to get crawfish." A mulberry-blue crawfish came into the boat from a cage that was deep in the Red Eye Swamp.

Farther down the trap line, Bourque said, "Crawfish is something hard to understand. When it's muddier, they're hungrier. The water's not muddy enough out here." There was a time when that sort of thing was a fact of nature. Now, of course, he blamed the Corps. "I'd like more water," he continued. "A lot of times, they've got much more in the Mississippi than they can use. They say they give us thirty per cent. We don't know if that's true."

I told him I had seen a tally sheet at Old River Control, and it said that 31.1 per cent had gone down the Atchafalaya the day before.

"I'd like to see that paper when the river starts dropping,"

Bourque responded. "I don't see that we get thirty per cent except when there is plenty of water. If they close the locks, it start dropping fast."

I mentioned the towboat Mississippi and its low-water Atchafalaya inspection trip, and asked if he had ever gone aboard to complain.

"I never heard of that until you mentioned it right now," he said. "They know we want more water. They don't have to ask."

I remembered Rabalais saying, "After they built the structure and started stabilizing this water and so on, the main complaint was the people from the Atchafalaya Basin—all your crawfish fishermen, and so on. They claimed they wasn't getting enough water, but over the years they've learned to live with it, and they catch as many crawfish, I would say, now as they did then."

And Peck Oubre, the lock mechanic, asking Rabalais, "Before they put in Old River Lock and the control structure, what was the people talking about when the water used to rise and come through here? Were they complaining about it?"

"No," said Rabalais. "They wouldn't complain, because there wasn't nothing you could do."

Bourque said that farmers who raise crawfish in artificial ponds—a fairly new and rapidly expanding industry—were influencing the Corps to keep the water low in the Atchafalaya in order to squeeze out swamp fishermen like him, whose forebears were swamp fishermen. It is possible that the charge he was making was based on pure suspicion, but now that the structures were emplaced at Old River—and the Corps had assumed charge of the latitude flow—suspicion was one more force they had to try to control.

As we were heading back toward the landing, Bourque remarked, surprisingly, "It's good we have the levees. Before the levees, the crawfish, they was spread all over."

For bait, for gasoline, and so forth, the cost of the day's run was seventy-five dollars. At the boat landing, Bourque sold the crawfish for three hundred and sixty. The buyer was Michael Williams, a youth from New Iberia with a mane of Etruscan hair. He identified himself as a poet, and said, "For poems there's not a market anymore. The days of the Romantic poets is gone. That's like in the past." So he also writes country-and-western lyrics. He recited one that began, "Oh, it's hard to write a love song/If you've never been in love." He had a pit bull named Demon with him. Demon went into the water and snapped at waves. He tried to bite motorboat waves.

I EMERGE FROM my remembrances standing at the rail, bewitched by the impenetrable vegetation. No part of those scenes that lie behind it can be felt or sensed from the decks of the Mississippi as the towboat moves on between the curtains of willow and straight down the middle of the bifurcated swamp. The others continue to talk, argue. The point is made that if the Mississippi River were to shift into the Atchafalaya the entire basin would fill with sediment and become a bottomland hardwood forest. "When nature shifts, man shifts," Oliver Houck says. The petrochemical industries would move to the basin, too, rebuilding themselves on Bayou Eugene, extruding plastics in the Red Eye Swamp. There are people in Morgan City who envision another Ruhr Valley up the Atchafalaya. Morgan City would be the new New Orleans.

The new New Orleans—seventeen miles from the Gulf—is not far ahead of us now. The landscape is changing to coastal marsh. Going below, I make a circumspect visit to the card game in the lounge. The Pontchartrain Levee Board draws three, Teche-Vermilion needs two. Ed Kyle, of Morgan City, whose pockets are familiar with United States currency bearing portraits that most people in their lifetimes never see and do not even know exist, throws one dollar into the pot. In the center of the table, the greenbacks reach flood stage.

Now, through the picture windows at the front of the lounge, our destination is in view: Morgan City, the Cajun Carcassonne—a very small town behind a very high wall. A railroad bridge and two highway bridges leap the Atchafalaya and seem to touch gingerly on the two sides, as if they were landing on lily pads. Flood stage in Morgan City is four feet above sea level. A dirt levee protected the town until 1937. It was succeeded by concrete walls six and then eight feet high. As floods grew—and the Atchafalaya became the only distributary of the Mississippi—sandbags and wooden baffles were piled up in haste on top of the eight-foot walls. Since it is the Corps' intention that fifty per cent of a Design Flood go down the Atchafalaya, and since Morgan City is on a small island of no relief situated directly in the path of the planned deluge, the Corps has built the present wall twenty-two feet high. It is of such regal and formidable demeanor that it attracts tourists. It is a wall that imagines water—a sheet of water at least twenty feet thick between Morgan City and the horizon. The seawall, as it is known, rises to the skirts of palms that stand in rows behind it. From the approaching towboat we can see a steeple, a flagpole, a water tower, but not the town's low avenues or deeply shaded streets. Damocles would not have been so lonely had he lived in Morgan City. In a proportion inverse to the seawall's great size, the seawall betokens a vulnerability the like of which is hard to find so far from a volcano.

Water approaches Morgan City from every side. The Atchafalaya River and its surrounding floodway come down from the north and pass the western edge of town. The seawall is a part of the floodway's eastern guide levee. When there are heavy local rains, as there were at the time of the great flood of 1973, water that is kept out of the floodway by the seventy-five miles of the eastern guide levee—water that used to go into the swamp and the river when the basin was under the control of nature—pools against the levee, caroms in the direction of the Gulf, and assaults Morgan City from the back side. The levee ends on Avoca Island, five or six miles south. The Atchafalaya floodwaters are sometimes so high that they go around the end of the levee and come back against Morgan City. Hurricanes also bring floods from that direction, surging from the Gulf like tidal waves.

Professor Kazmann, of L.S.U., said, "You can't sell Morgan City short, or I would." To end its days, Morgan City does not require a Design Flood. The Design Flood, at Morgan City, is a million and a half cubic feet per second. LeRoy Dugas, of Old River, once explained to me, "The Old River Control Structures can pass seven hundred and fifty thousand cubic feet per second and the Morganza Spillway six hundred. In that situation, if both of them are wide open, we've got Morgan City gasping for air." The people of Morgan City are not easily frightened. They would tell Professor Kazmann to get back into his college and Dugie to shut a few gates. Mayor Cedric LaFleur says, "I feel safe. I feel secure. We're not going to wash away." If there is a slightly hollow sound as he speaks, it is because Morgan City is sort of like a large tumbler glued to the bottom of an aquarium. The Corps, of course, built Morgan City's great rampart, and graced it with bas-reliefs of shrimp boats and oil rigs—consecutive emblems of Morgan

City booms. Everyone is grateful for the wall. Morgan City—
in its unusual setting—is dependent on the Corps of Engineers
in the way that a space platform would depend on Mission
Control. The fate of Morgan City is written at Old River.
Anything that happens there is relevant to the town.

As the towboat passes under the second bridge and turns
toward a berth below the seawall, I ask General Sands what
sort of complaint he most frequently receives when he comes
here. He says, "The Corps of Engineers isn't doing enough
to protect Morgan City from disaster."

The hearing is at nine the next morning, aboard the
Mississippi in the thoroughly transformed lounge. Where
Teche-Vermilion was taking pots, the scene is now set for the
court-martial of Billy Mitchell. In front of various standing
flags, the three generals and two civilian members of the Mississippi River Commission sit at a large formal table, with
General Sands in the central position. A colonel is master of
ceremonies, and three other colonels are in the front row. This
seems an unlikely place for Clifton Aucoin to present his
petitions, but now he stands before them—a man in bluejeans
and an open shirt, whose remarks suggest that he has spent a
good many days of his life up to his hips in water. "My name
is Clifton Aucoin," he testifies. "Very few people pronounce
it right, so don't feel bad about it." He tells the commission
that he once kept a boat tied to the knob of his front door.
"As far as us people in the back floodwater area, we feel ne-
glected," he continues. "As far as we can tell, nothing has
been fixed. Atchafalaya water just comes around Bayou
Chene, it comes right on us backwater people. . . . We feel
that it's just another major flood that's waiting to hit us if
nothing is done about it." As a hunter, he further complains
of dying trees, of disappearing browse and cover—changes no

longer ascribable to nature but now quite obviously conceded to be under the control of the Corps.

The commissioners hear Cedric LaFleur, a trimly built man with curly hair and dark, quick eyes. LaFleur says it is "a dire relief" to have the seawall completed, and suggests that the Corps stop studying the Avoca Island levee and extend it several miles south—to prevent the floods of the Atchafalaya from going around the levee's tip and coming back upon the town. Terrebonne Parish, east of the proposed extension, has complained to the Corps that an extended levee would deprive Terrebonne marshes of sediment, thereby destroying the marshes. The survival of one parish is in conflict with the survival of another, and each is appealing to the Corps.

They hear Mark Denham, of St. Mary Parish: "We appreciate y'all coming down. We really consider having the Corps as a presence in our area a tremendous asset to our area as far as protection of floodwaters and as far as economic development also."

They hear Jesse Fontenot, Curtis Patterson, Gerald Dyson—chambers of commerce, levee boards, the government of the state. And, as they inevitably do in Morgan City, they hear Doc Brownell. He comes forward slowly, slightly stoop-shouldered, septuagenarian. This man once entered prize-fights. There is a trace of smile on his face. He, too, thanks the commission. "It's always a pleasure to see you people come down here. It gives us a little encouragement." And then, in effect, he tells the Corps to get its act mobilized and extend the levee. For thirty-two and a half years, Doc Brownell was the mayor of Morgan City. LaFleur has been described as his
_____ hen the water went around the end of the
ack up Bayou Chene, Brownell, without
teen-hundred-ton barge in the bayou. The

barge acted as a dam and held off the water long enough for the people to build up their defenses and save the city. "The nightmare of '73 is still with us," Brownell reminds the commission. "We live in a state of apprehension; we live on the whims of the weather of over forty-two per cent of the United States. . . . We live with it twenty-four hours a day." He praises the beauty of the new seawall but points out that to the people of Morgan City its extraordinary height is an unambiguous message from the Corps. "We can expect that much more water. It makes us very apprehensive. We have got to extend our defenses."

Brownell, who went into medicine because the lumber business was dying, became a sort of bayou Schweitzer, delivering babies far out in the swamps, doing surgery in an unair-conditioned operating room for twelve and fourteen hours a day. Among his closest companions was an alligator called Old Bull, who lived with the Brownell family for thirty-five years. Old Bull died in 1982 and is now in a glass-sided mahogany-framed case—in effect, a see-in coffin—looking almost alive among simulated hyacinths, iris, and moss in Brownell's parlor. Tip to tip, Old Bull is ten and a half feet long. There is a brass footrail next to Old Bull and a padded bar above him, with beer tap, soda siphon, and a generous stock of bottles. Brownell took Charlie Fryling and me there one spring day to admire Old Bull and to show us, with the help of pictures, the predicament of Morgan City. What struck me most of all as he talked was his evident and inherent conviction that a community can have a right to exist—to rise, expand, and prosper—in the middle of one of the most theatrically inundated floodplains in the world. To be sure, the natural floodplain is also an artificial floodway—concentrated and shaped—and, accordingly, its high waters are all the more severe. In Morgan City, it has become impossible to separate the works of people from the periodic acts of God. "We have a lot of restaurants now and various types of establishments in places vulnerable to the water," Brownell said. "We got to develop on the floodplain. It's the only place we got to develop. We still have got to look for places for people to live. Now, you can see from this map that we're right in the middle of this floodway. It's like a funnel with a spout, and we're at the end of that spout. We're in the concentration part of it. We have our homes, our families, our whole future in the floodway. We've got to live with these problems—and to me it ought to be some type of priority for the people who live under these conditions twelve months out of the year should be given some type of preference as to what our future is. It's the nation's problem, and we are only the victims here of a lot of things that does happen here that are imposed upon us. We lost the big live oaks in the park because of the long-standing floodwater. A flood doesn't last for weeks here, as it does in some of those northern places. Our floods last for months. The more ring levees are built to the north, the more water Morgan City gets. In whatever way the people upriver protect themselves, they send more water to Morgan City. If people dig canals to get water off their land, it goes to Morgan City. When you're drowning, you don't need more water."

Tarzan of the Apes once leaped about among the live oaks in the park. The first Tarzan movie was filmed in Morgan City. The Atchafalaya swamp was Tarzan's jungle. Black extras in costumes pretended they were Africans.

Not far from Old Bull, the head of another alligator was
—its mouth open, a light bulb in the back
ed owls and hawks were hanging on the
geese were flying through the air. There

(84

were the heads of deer, of black bears from the Atchafalaya swamp. Brownell said his father had killed six bears shortly before he died. There was a stuffed tarpon head as large as the head of a horse. The tarpon was caught in the Atchafalaya River near Morgan City before the river, increasing in volume and power, pushed back the salt water. Islands now stand where the river was a hundred feet deep. As the Atchafalaya has grown, more and more sediments have, of course, come with it, stopping where they reach still water. This is the one place in Louisiana, other than the mouth of the Mississippi, where new coastal land is forming. Large areas of what was once Atchafalaya Bay have become dry flats. The soil broke the surface as the flood receded in 1973. Whole islands appeared at once. The bay was choked. Brownell says the river built a dam there. A geologist would call it a delta.

Charles Morgan, a shipper in New Orleans in the eighteen-fifties and sixties, was so irritated by New Orleans' taxes, New Orleans' dockage fees, and New Orleans' waterfront clutter that he moved his operation to the Atchafalaya and developed a competing city. It seems unlikely that he was aware that the Mississippi River meant to follow him. Morgan City thrived on shipping, on oysters. When the big cypresses were felled in the Atchafalaya swamp, Morgan City became the center of the cypress industry in the United States: numerous sawmills, hundreds of schooners in the port. Brownell's great-grandfather owned a sawmill. In the nineteen-thirties, Captain Ted Anderson, a Florida-based fisherman, was blown off course by a storm, and put in at Morgan City. In the hold of his boat were shrimp of a size unfamiliar in Morgan City— big ones, like croissants, from far offshore. They were considered repulsive, and at first no one wanted them, but these jumbos of the deep Gulf soon gave Morgan City the foremost shrimp fleet in the world. As the Atchafalaya River pushed back the salt water, it pushed out of the marshes the nurseries of shrimp. Caught in the westbound littoral drift, the shrimp went to Texas, where much of the business is now. The growth of cypresses was too slow to keep up with the lumber industry, so the lumber industry collapsed. The next boom was in oil. The big offshore towers come out of the marshlands surrounding Morgan City. They are built on their sides and dominate the horizon like skeletons of trapezoidal blimps. Of the twelve hundred and sixty-three permanent platforms now standing in the Gulf on the continental shelf, eighty-eight per cent are off Louisiana.

In other words, the people of Morgan City are accustomed to taking nature as it comes. Cindy Thibodaux, the town archivist—a robust young poet with cerulean eyes and a fervent manner of speaking—said to me one day, "When you're fishing in the bayou, you're out in nature with the oil industry all around you." She has written a poem about the oil industry and nature from an alligator's perspective.

IN THE PRESENCE of the tribunes on the towboat, as the Pontchartrain Levee District recites its needs and the State of Louisiana its concerns—as the discussion touches upon the varied supplications of the whole deltaic plain, and on the growth of the extremities of the great levee system not only below Morgan City but down the Mississippi from Bohemia to Baptiste Collette—my mind cannot help drifting back to Old River, where every part of this story in a sense had its beginnings and could also have its end. Near the mouths of the intake channels of Old River Control, the Corps maintains another towboat, smaller than the Mississippi but no less pow-

erful—a vessel on duty twenty-four hours a day and not
equipped with white couches, wall-to-wall windows, or vene-
tian blinds—the name of which is Kent.

Kent is a picket boat. It defends Old River Control. With
its squared bow and severed aspect, it appears to be a piece of
wharf that loosened like a tooth and came out on the river.
Kent's job is to catch, hold, and assist any vessel in trouble.
If barges break loose upstream and there is insufficient time
to tie them up, Kent is supposed to divert them. Technically,
it is a twin-screw steel motor tug, eighty-five feet long, with
two nine-hundred-horse diesels that can start at the touch of
buttons. (Compressed air makes that possible.) It cost two million dollars and differs from most river towboats only in its
uncommon electronics—the state and variety of its radar, the
applications of its multiple computers. In addition to the on-
board radar, two radar beams sweep the river from the bank
at stations four miles apart, and anything that reflects from
these beams appears on a screen in Kent. If a tow rig is moving
at the speed of the current, an alarm goes off, for the coin-
cidental speed suggests that the rig is without power. Kent can
tell this eight miles away.

Fifteen miles up the river, in April of 1964, twenty barges
full of ore were tied to the bank and left there unattended.
Eight of them broke free. There was no picket boat then. As
a functioning valve, the control structure at Old River was
nine months old. As the ore-laden barges drifted near, they
were drawn away from the Mississippi, sucked into the structure by the power of the Atchafalaya. One of them plunged
through the gates and sank on the lower side. Three sank in
front of the gates and effectively closed the structure. A standard barge is a hundred and ninety-five feet long. Water piled
up. Weeks went by. Much of the time, the difference in water

level between the Mississippi and Atchafalaya sides was thirty-five feet, a critical number that resulted in damage and "threatened the integrity of the structure"—the Corps' way of saying that it might have been wiped out.

Today, it is illegal to tie anything to either bank of the Mississippi within twenty upstream miles of the structures at Old River. Every approaching vessel has to radio Kent and, as Dugas puts it, "say what he is, who he is, and if he has a red-flag product." And for ignorant river pilots and all uninitiated craft there's a very large sign high up the bank of the river—its first three words in red:

WARNING
DANGEROUS DRAW
1 MILE—WEST BANK
OLD RIVER CONTROL STRUCTURE
U.S. ARMY
CORPS OF ENGINEERS
NEW ORLEANS DISTRICT

Spring high water often knocks the sign away.

It would be difficult to overestimate the power of the draw, deriving, as it does, from the Atchafalaya, by now, in point of discharge, one of the twenty to thirty strongest rivers in the world. The Coast Guard once tried to set five warning buoys in the west side of the Mississippi, but could not keep them in place, because the suction was so fierce. This threat to navigation could be called an American Maelstrom—a modern Charybdis, a Corryvreckan—were it not so very much greater in destructive force. In Dugie's words, "Any rig on the right side of the river is in trouble."

An empty barge and three barges loaded with quarry stones were sucked into the low sill in 1965. Two loaded barges went through the structure and sank on the Atchafalaya side. The other sank against the gates without causing apparent damage, but it must have contributed to the turbulences that even then were undermining the structure. After the great flood of 1973 and the considerable debilitation it disclosed, there was the constant danger that if several loose barges were to block the flow and the difference in water levels were to build to catastrophic proportions nothing could be done about it. One barge spent a flood against the gates in 1974, but the structure survived.

People in Simmesport often refer to Old River Control as "the second locks." John Hughes, the supervisor of Kent and one of its operators, does his best to correct them. "That's not a lock, that's a control structure," he says. And a Simmesport person says, "Well, we was born and raised here, and we call it the second locks." To judge by the amount of traffic erroneously attracted to the control structure, they have a point. A boat comes down the river, takes a right, and heads for Old River Control, thinking that it is Old River Navigation Lock. Usually, the boat is small—a cabin cruiser, or something of the sort—but the mistake has been made by a fifteen-barge tow. Its skipper called in on the radio to the navigation lock, announcing his arrival. The people at the lock replied that they didn't see him. He said, "I'm right here looking at you, I'm coming in." The mistake was corrected just in time.

In 1982, thirty-nine barges broke loose thirteen miles upstream at four in the morning. The whole rig just came apart. Dugie recalls, "He was in a bend of the river. He couldn't maneuver the river. He hit the bank." The picket boat went after the barges. Five other skippers, joining their units together, detached four towboats that came to help. "They could see the picket boat had a lot of problems, trying to catch thirty-nine barges by hisself," Dugie says. At 6 A.M., right at the entrance to the intake channel of Old River Control, the last barge was caught. Not even one hit the gates. Two of the thirty-nine were red-flag barges, loaded with petroleum. Later that year, a fifteen-barge rig heading north in the dark swung too close to Old River Control, was drawn off course, and—its engines overmatched by the force of the water—crashed in the sand on the north side of the intake-channel mouth. In 1983, at midnight, a towboat with three jumbo barges lost power at Black Hawk Point, two miles above the structure. The picket boat caught it before it reached the channel.

The operator on that occasion was Gerald Gillis, whose broad full face and long jet-black hair lend him the look of an Elizabethan page after twenty-five years in Morgan City. He is one of eight men who work Kent—two on a shift. One day, he took me out on the beat with him, running up the river. He said the speed of the Mississippi current ranges from about three knots in low water to six in spring and eight in flood. A rig coming downstream on this September day would be averaging about eight knots. To conserve fuel, the big thirty-five-barge tows like to crawl along just barely ahead of the speed of the river, and that confuses Kent, because the tows could be dead in the water. An example was descending toward us now, called Gale C, shoving thirty-five barges of grain and coal, and much alive in the river, as Gillis learned from his transceiver. While the huge rig was passing by us—really an itinerant island, eight thousand horsepower and a quarter of a mile long, with its barges in seven ranks of five—he said the rough rule of thumb for fuelling such an enterprise on journeys upstream is one gallon per horsepower per day.

Gillis turned on the depth finder. We had come up the

Mississippi's east side, and now he swung crosscurrent, heading for the cutbank of the west-convexing bend just above the structures of Old River. As we traversed the Mississippi, the depth, which was being sketched by a stylus on graph paper, dropped steadily and kept on dropping the closer we came to the bank. We were only a few swimming strokes from shore when the depth reached a hundred feet. It was notable that the riverbed was fifty feet below sea level more than three hundred miles from the mouth of the river, but what particularly astounded me was the very great depth so close to the west bank. It showed the excavating force of a tremendous river. The foundations of skyscrapers are rarely that deep. And this was the bend where the water swung off and into Old River Control—a bend armored with concrete where the Mississippi might break free and go to the Atchafalaya. Kent was so close to the bank that it had no room to turn. Gillis backed away.

Twenty years before, a barge that broke loose and was crumpled after sinking at the structure was hauled up the intake channel and left by the edge of the river. The barge had not moved since then, but the Mississippi's bank—consumed by the scouring currents—had eroded to the west. The barge now lay five hundred feet out in the Mississippi.

General Sands, reflecting on these matters, once said, "The Old River Control Structure was put in the wrong place. It was designed to a dollar figure."

And Fred Bayley, his chief engineer, added, "That is correct. It was done during the Eisenhower Administration."

The Corps once attempted to barricade the intake channel with a string of barges anchored in the river. Drift—as the big logs are called that unremittingly come down the river— amassed against the anchoring cables until enough had gathered to heave high and start breaking the cables. As if drift were not enough of a problem, ice has been known to appear as well. It may come only once in twenty years, but ice it is, in Louisiana.

The water attacking Old River Control is of course continuous, working, in different ways, from both sides. In 1986, one of the low-sill structure's eleven gates was seriously damaged by the ever-pounding river. Another gate lost its guiding rail. When I asked Fred Smith, the district geologist, if he thought it inevitable that the Mississippi would succeed in swinging its channel west, he said, "Personally, I think it might. Yes. That's not the Corps' position, though. We'll try to keep it where it is, for economic reasons. If the right circumstances are all put together (huge rainfall, a large snowmelt), there's a very definite possibility that the river would divert—go down through the Atchafalaya Basin. So far, we have been able to alleviate those problems."

Significant thanks to Kent.

A skiff rides on Kent's stern. A part of the skiff's permanent equipment is a fifteen-foot bamboo pole. Kent is alert to everything that moves in the river, including catfish.

Cooling the Lava

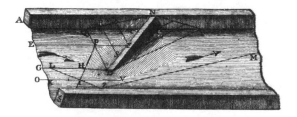

COOLING THE LAVA was Thorbjorn's idea. He meant to stop the lava. That such a feat had not been tried, let alone accomplished, in the known history of the world did not burden Thorbjorn, who had reason to believe it could be done.

His full name is Thorbjorn Sigurgeirsson. If you look for him in Simaskra, the Iceland telephone directory, you look under "Thorbjorn." You look under "Sigurdur" for Sigurdur Jonsson. You look under "Magnus" for Magnus Magnusson. I had occasion recently to call all three of these, and finding them in Simaskra was not for me a simple task. It doesn't matter that Sigurdur is a harbor manager's deputy, or that Magnus is a postal director, or that Thorbjorn is a physicist trained in Copenhagen by Niels Bohr. Like the Prime Minister, like the President—like all people in Iceland—Thorbjorn is known by his first name. Sigurgeir, of course, was his father.

To Thorbjorn's idea skepticism was the primary response. The skeptics included Magnus Magnusson, Sigurdur Jonsson, Valdimar Jonsson, Thorleifur Einarsson, Gudmundur Karlsson, and almost everybody else in Iceland. Red-hot lava—moving with the inexorability of tide—was threatening a town and a harbor on an offshore island. The vent was a nascent volcano. As the entire nation watched on television, a small crew with fire hoses squirted the front of the lava, producing billows of steam. This was in February, 1973. Quickly, the cooling of the lava became a national joke. The people called the action pissa a hraunid, in which "a hraunid" meant "on the lava."

There may even have been smiles in the National Emergency Operation Center, in Reykjavik, where the Civil Defense Council was watching Thorbjorn's experiments. The National Emergency Operation Center has grown since then but is still concentrated around a command-post bomb shelter built as Iceland's cerebral cortex in the event of a nuclear war. The Civil Defense Council had been established in 1962, and was scarcely a year old when—sixty-five miles southeast of the capital—the ocean began to boil. The Council became preoccupied with forces greater than bombs. Red lava appeared in the Atlantic swells and, layer upon layer, emerged as an island that was soon the second largest of a fleet of islands collectively known as Vestmannaeyjar. The fresh lavas of Surtsey, as the new one was called, hissed into the ocean for three and a half years. In the same period, unusual masses of drift ice blockaded various harbors, causing wide destruction; and the charter purpose of the Civil Defense Council was broadened to include acts of God. Behind thick concrete walls and steel anti-radiation doors, in a space where the air can be pressurized to keep anything noxious from seeping in, Iceland directs the war against nature. "War" is the word often used, especially with reference to the campaign begun after the twenty-third of January, 1973, when a fissure suddenly opened in the outskirts of a community of five thousand people and a curtain of lava five hundred feet high and a mile long fountained into the sky above Heimaey, the largest island of Vestmannaeyjar. All these years later, the communication room of the National Emergency Operation Center, in Reykjavik, is decorated with a relief model of Heimaey and eight large photographs, ruddy with violence, made in that place and time, for, notwithstanding the frequency of eruptions in various parts of Iceland and of natural disasters in many forms (sea surges, earthquakes, glacial bursts), Heimaey is the most signal battle ever waged by the Civil Defense.

Heimaey is pronounced "hay may." Vestmannaeyjar is more or less pronounced "vestman air." The town on Heimaey is the only place in the archipelago inhabited by human beings, and it has no special name. It is referred to simply as Vest-mannaeyjar or, less frequently, as Heimaey, which means "home island." In area, the home island is so small that it approximates Manhattan south of the Empire State Building. The volume of material that came pouring out on Heimaey in 1973 would be enough to envelop New York's entire financial district, with only the tops of the World Trade Center sticking out like ski huts. The image is not as outlandish as it seems. A few miles west of Manhattan, the high ground of Montclair—of Glen Ridge, Great Notch, and Mountainside—is the product of a similar fissure eruption. In Vestmannaeyjar, the nub of the crisis was simple and economic. Proportionally, Heimaey was more valuable to Iceland than downtown New York is to us. Vestmannaeyjar, with two and a half per cent of Iceland's people, was producing about a twelfth of Iceland's export income. Vestmannaeyjar was a place to catch fish. Iceland's exports were three-quarters fish. As a result of Heimaey's natural protection, it was Iceland's single most important fishing center. Not merely was Heimaey's harbor the best harbor along the three hundred miles of Iceland's south coast; it was the only one. That is why this eruption, as someone said, "just came like a thundering cloud over the whole nation." That is why there was so much alarm when a generally slow and viscous, magisterial lava—dark-shelled by day and a craquelure of red and black at night—began to move not only toward sections of the town but, more important, toward the entrance to the harbor.

The cooling was at first confined to the lava front. Men stood on cold ground before the flowing rock and watered it like a garden. Its Fahrenheit temperature was around two thousand degrees. The water would reduce the heat locally by a factor of four, creating a wall of chilled lava to dam the flow behind. Soon it became apparent that the wall would have to be a great deal thicker than hoses could ever make it from positions ahead of the flow. The lava should be cooled not so much by the edge as by the acre, and that called not only for more pumps but also for the deployment of matériel and personnel up on top of the advancing flow.

It was astonishing to see what an essentially liquid body of rock would carry on its surface. As lava moves, under the air, it develops a skin of glass that is broken and rebroken by the motion of the liquid below, so that it clinks and tinkles, and crackles like a campfire, which, in a fantastic sense, it resembles. In Hawaii, I have been close enough to the flowing lava of Kilauea to ladle some out and carry off a new rock, but when I visited Vestmannaeyjar the eruption had been over for a decade and a half, and, needless to say, no lava was moving. Steam was still rising from the lava field, though, and grew to heavy mist under rain. If you reached down and put a hand on the ground—on loose ash—the new surface felt cool. If you rubbed away as little as a third of an inch, the ground was so hot you had to pull back your hand. The ash was such an effective insulator—not to mention the layers of solidified basalt underneath it—that the interior of the 1973 flow was still molten. During the eruption, when the pumping crews first tried to get up onto the lava they found that a crust as thin as two inches was enough to support a person and also provide insulation from the heat—just a couple of inches of hard rock resting like pond ice upon the molten fathoms. As the crews hauled and heaved at hoses, nozzle tripods, and sections of pipe, they learned that it was best not to stand still. Often, they marched in place. Even so, their boots sometimes burst into flame.

It was soon apparent that bulldozers were required on the lava—to flatten the apalhraun (the jagged surface glass, known elsewhere in the world by the Hawaiian word "aa"), to make roads for the pumping crews, to move some very heavy pipe. With the eruption going at full volume, bulldozers crawled up onto the flow. About a foot and a half of crust was enough to support them—to keep them from plunging through. The bulldozers worked in dense warm fog—the remains of the seawater pumped on the lava. The fog was so thick that a bulldozer operator could not see his own blade. Sigurdur Jonsson, who had lived all his life on Heimaey and at the time of the eruption had been working as a clerk in a hardware store, served from the outset with the pumping crews, and at times he was assigned to walk beside bulldozers and help guide them. Five people worked each dozer, he told me—the operator up in his cab, a person at either end of the blade, and the two others at the sides. The steel tracks of the big machines became so hot that they turned dark blue.

"One time, a bulldozer caught fire. Thirty metres away

was a little pool where the water line had broken. He went into the pool. He came out and went to work again."

Sigurdur wore oversize boots with two or three pairs of wool socks.

"You couldn't stand for thirty seconds. In some areas, the lava was glowing. You would see a glow here, a glow there."

Not much below fifteen hundred degrees, lava stops glowing. Even if you can see it, you may not sense how hot it is.

"Leather boots shrank under the heat—Icelandic leather boots with lambskin, made for extreme cold. People carried water bottles to pour water into their shoes. We wore ski goggles, plastic goggles. Often, we could not see. You could not see an arm length. You found your way by the noise of the volcano, by the shape of the lava, by the pipelines. The pipelines guided us backward and forward. You used the pipeline all the way, working on the lava. Sometimes you saw the light of the sun through the fog. As the cooled area was getting bigger, visibility improved."

When the fog temporarily cleared, it was a straight shot of twenty miles across open water and up coastal slopes to the white walls of Eyjafjallajokull—at five thousand feet, a frozen cloud. A tenth of Iceland is capped with ice.

When Sigurdur worked at the harborside, spraying the advancing front, winter winds were fiercely cold.

"You were in the sea spray. You got ice on your shoulders and chest. You went up on the lava and the ice would melt."

One day, seaweed plugged an intake for the pumps. He dived into the water and cleared away the seaweed.

The pumping platoon, at maximum, numbered seventy-five people, in a corps of five hundred who were engaged in the battle at any given time. Many came from Iceland (as the mainland is known on the island) to serve forty-eight hours at a stretch, and then go home. Sigurdur never left the island. He was, among other things, the head of the local chapter of Hjalparsveit skata—one of a number of first-aid-and-rescue groups, which are the closest thing to a militia known in Iceland—but he was not a leader in the fight against the lava. He was a foot soldier, and looked upon his role as such. Like the others, he viewed the struggle through the metaphor of war. "When the steam stops coming, and you see water running down the lava, you move the hose two or three metres. The lava front is black. Suddenly you get the red through the black. It could take hours—even days—to make it black. Sometimes it didn't work. You had to withdraw. You retreated, retreated until you were too close to the supply line; then you moved the supply line. In the battle, if you did not have to withdraw hoses and pipelines that was a victory for the day."

People behaved as if they were in combat, and on the streets of Reykjavik and other Icelandic towns veterans encountering one another will still talk about their service in Vestmannaeyjar.

The vapors of the eruption affected people's throats. Mouths were covered with cloth. "What the steam did to your throat—it made everybody hoarse," Sigurdur told me. "In the early days of the volcano, we were more hoarse. The doctors gave us medicine for it, which included Norwegian chest drops. You know what I mean by Norwegian chest drops."

"Ninety proof."

These recollections took place one summer evening in Sigurdur's house, Hasteinsvegi 47, in a close-set row on a tranquil street. Gentle and quiet in manner, a somewhat heavy man, partly bald now, with a mustache and gold-rimmed glasses, Sigurdur apologized frequently for his English, which he spoke at about the level of the average American professor. He had offered me a Heineken—a particularly generous gesture, but not so much of one that it would cause me to refuse. In Iceland, Norwegian chest drops—all the high hard ones from Jonny Walkersson to Jack Danielsson to Lambeater gin— are legal, and are sold over the counter at government stores. Beer, however, is prohibited, in the interest of the national health. The only legal beer brewed in Iceland is virtually nonalcoholic. In a somewhat paradoxical country, this is a savory paradox. Icelanders go to public bars that scrupulously obey the law. Not as a result of disrespect is theirs the oldest sitting democratic parliament in the world. Across the bar comes a bottle of prescription beer, accompanied routinely by a throat-burning shot of schnapps.

My Heineken in hand, I thanked Sigurdur warmly, and said, "It's hard to come by in Iceland."

He said, "It depends on what you do."

Continuing his narrative, he remarked that the eruption—for all its great surprise and early spectacle—had grown slowly.

"We had time to get used to it. You tried to protect your feet. I lost boots—I don't know how many. Only once I had to go to the doctor. Also your wrists. If you fell down, the wrist would open between gloves and coat. So we wrapped our wrists with bandages. Very soon we stopped using helmets. Steam got inside helmets and hot water dripped on your head. We got from people's houses old gentlemen's hats with brims. They protected our faces. We had American Army helmets. We could always reach for them."

The helmets were for protection against falling ash and, in proximity to the crater, huge volcanic bombs. The ash was typically about as large as bits of pea gravel. They fell by the millions—hot enough and sharp enough to burn and cut skin. The bombs were ejections of molten lava that flew high into the air, became spherical, and, with contained gases, sometimes exploded like fireworks. In their interiors, bombs were generally molten. After they landed and broke, red-hot liquid poured out. Bombs that landed as much as two-thirds of a mile from the crater might weigh as much as sixty pounds. Bombs that carried half that distance could weigh a third of a ton. Sigurdur Jonsson and the others in the pumping crews usually worked outside the radius of the big bombs—except when they were on crater watch, in an ironclad hut quite close to the volcano, where forward observers could see into the churning lava, report its current style of eruption, and warn their colleagues in the fog about fresh eruptive flows.

I asked him, "What if a big fiery rock came down when you were working blind like that on the lava in the dense fog?"

He said, "It just didn't."

IF A CREWMAN was in clear air with bombs above, there were ways to avoid the bombs. Thorbjorn taught recruits his own technique. "When you come to a crater and it starts blasting, don't run," he advised them. "Look up in the air for a bomb. If you think one is coming down on you, wait until you are sure it will hit you, then step aside." People walked in pairs—one leading the way with eyes on the ground, the other looking up.

Thorbjorn said "a crater" because he had been close to others, equally active—notably on Surtsey. During the long Surtsey eruption, he went there routinely to record temperatures and measure the magnetic field. This was not a typical destiny for someone schooled on fission tracks and particle physics—for a scientist who had moved on from Copenhagen to Princeton, had done research on cosmic rays with John Archibald Wheeler, and had designed an experiment that measured the life of the shortest-lived particle then known. Thorbjorn, however, had returned to Iceland. In Iceland, as one of his colleagues has put it, "atomic energy is neither interesting nor important," and, *de rerum natura*, almost every scientist of any description sooner or later turns into a volcanologist. The colleague continued, "That is something that the environment does to Icelanders. Physicists, astronomers come home—they have no cyclotron, no observatory. The land just calls them, and they go into geophysics."

"You cannot avoid that when you live here," Thorbjorn said when I called on him in his office at the University of Iceland. He is slim in build, has a tall, thin face, a prominent nose. In his white turtleneck, he seemed to be a stretched George Washington. His hair, largely gray now, is closely trimmed at the sides, and sticks straight up in back in a manner that suggests astonishment. His smile is very large, and flashes often. He has been described as "a physicist of the screwdriver type, who can fashion his own instruments and make do with simple things." When he became interested in paleomagnetism, in the middle nineteen-fifties, he developed an airborne proton-precession magnetometer, learned to fly, and flew all over Iceland. On Surtsey in the sixties, he told me, a hut was threatened by flowing lava. Watching other lava approach the sea there, he had noticed that it flowed to the beach and then followed the coastline for a long distance. "The sea cooled it," he explained. "Then lava ran along that cooled wall. I wondered, could anything similar be done by man?"

To experiment, Thorbjorn obtained some fairly big

pumps, which, in 1966, a ship carried to Vestmannaeyjar, where it waited in the harbor, in bad weather. In Vestmannaeyjar, as in much of Iceland, it is not easy to differentiate between bad weather and a natural disaster, because hundred-knot winds are common and waves will break over an islet two hundred feet high. In any case, the storm was in no hurry to abate, and the ship with the pumps went back to Iceland. Eventually, Thorbjorn tried some small pumps on Surtsey, but, he said, "it was not a real test."

I asked what had drawn him to Heimaey, where a helicopter set him down a few hours after the eruption began.

"Curiosity," he said.

But it was more than that. Icelandic scientists had conferred at once about who should do what. Thorbjorn was a former director of the National Research Council of Iceland—the provost of all Icelandic science. He went to Heimaey specifically to measure lava temperatures, but mainly to be on the scene. He saw the classic pattern the eruption was following. The lava curtain drew in from the two ends and began to concentrate in the place where the volcano would grow. Lava poured from this central vent and fanned eastward to the sea and north toward the harbor. Thorbjorn's living quarters were in the aquarium (he calls it "the natural museum"), close to the center of town. The Heimaey fire brigade was on the ground floor of the building. Thorbjorn looked over the fire engines, which were actually small vans equipped with pumps and hoses. They were kept busy fighting house fires ignited by falling bombs. But perhaps not too busy. Against the basso profundo of the volcano, the sporadic explosions, and the sound of raining ash, he had talked with the firemen and interested them in his idea. "I also wrote memoranda to the

Civil Defense Council," he said, finishing the story. "There was no immediate response. The actual response was through the fire people."

About a fortnight after the eruption began, Thorbjorn and his aquarium-mates went out for the first time to water the lava. They chose a position on the shoreline roughly a thousand feet outside the mouth of the harbor. The harbor was a long wedgelike cove, open to the east, and its mouth was defined by a pair of offset breakwaters reaching in from the two sides. As it happened, there could not have been a more public place in which to conduct the lava-cooling experiment. Across the harbor, behind the firemen and their hoses, was Storaklif—a flat-topped sea stack about seven hundred feet high, rising sheer from the water. On its summit was the apparatus that normally received television signals from the mainland and bent them down to the island. Now a television camera on top of Storaklif was reversing the process. Unblinking, never panning, it stared at the new volcano and the intervening lava. The scene was broadcast by Iceland's lone television station nineteen hours a day.

As the hoses made their trajectories against the dark advancing mass and the water blew away as steam, Thorbjorn, of course, could not hear the laughter. Thorbjorn could not see his colleague Valdimar K. Jonsson, Professor of Thermal Fluids—a mechanical engineer who would play a large role in the coming operation—watching its beginnings with incredulity in his living room on the mainland. ("I laughed. I laughed because I did not believe that we could stop Mother Nature. The television showed a small fire truck putting water up against the lava front. I laughed. I asked myself, 'What are they really trying to do?' ")

The volcano was shooting ash straight up five miles.

Where lava was entering the sea, bright white steam went five miles upward, too, joining en route the darker volcanic cloud. Who could not agree that the odds seemed to favor nature? Thorbjorn was impressed by the early results of his experiment. Where the hoses had played on the lava, its movement was arrested, while the unwatered flow on either side moved on, forming lobes—somewhat like breasts, somewhat like eyeglasses. After all the intervening years, Thorbjorn, recalling this, still seemed pleasantly surprised. "We were having some effect," he said. "It solidified. It stood still."

As the spreading lava advanced in the direction of the town, pumps obtained on the mainland drew water directly from the sea in an effort to save houses. Thorbjorn said of this second stage, "The edge of the lava became very steep and very high. Lava jets came out of the base. The idea was to be quick to cool them. The edge was never stationary. We could not cool it completely. It was very difficult to get hoses up on top. Some of the water was simply running down. Only a rather thin layer was being cooled. We needed to have water up on top."

Thorbjorn was later able to calculate that one cubic metre of water would change seven-tenths of a cubic metre of lava from incandescent flow to hard rock; and what began as the effluent of a few fire hoses would eventually amount to six million tons of water pumped on the lava—the equivalent of turning Niagara Falls onto the island for half an hour. In the first few weeks of the eruption, however, when Thorbjorn's experiments were only indicative, no one could have guessed the scale of the effort that was coming. Meanwhile, the height of the molten fountains reached seven hundred feet, and lightning flashed beside them. The sea boiled at the edge of the lava. When the crews went off exhausted after fourteen hours of work, they were at times unable to sleep, because, as Hawaiians would put it, they were "hearing Pele": buildings shook, windows rattled, and the inferno went on grumbling. For the most part, it was not explosive. Bjarni Sighvatsson, born on Heimaey and twenty-three at the time of the eruption, has told me, "It more felt than sounded. You felt it when you slept, not when you were working. The sound was like thunder, but distant thunder. The fire curtain also did not make much sound. It was like hearing, from a distance, waves hitting cliffs." When the eruption started, the wind had blown from the west for a time, pushing the fallout into the sea, but then the wind shifted and down upon the town came hundreds of tons of tephra—debris of all sizes from ash and gravel, through stones the size of apples, to bombs that broke windows and landed inside them. Houses burst into flame. Carpenters from the mainland, arriving by the platoon, began to cover windows with sheets of corrugated iron. In a week, they covered fourteen thousand windows. Volunteers from the mainland shovelled ash from rooftops. A small bulldozer was hoisted to the top of the hospital, where it pushed away tephra five feet deep.

A large part of these operations—including, eventually, the coordination of the pumping crews—was directed by the Icelandic fire chief of the American base at Keflavik, where fighter-interceptors are always warm. This was a slender man of deceptively mild aspect, vaguely professorial, appearing like a genie through his own pipe smoke. He sometimes wore a uniform, with stripes that suggested military rank, but he was an Icelander, not a soldier, and, in any case, no width or number of stripes could ever have conveyed the status he acquired on the island. Sent by the Civil Defense to help in the emergency, he quickly assumed command of one unit after another, until his de facto rank had outflown eagles and was far into the stars. His name was Sveinn Eiriksson, but no one much used it. On Heimaey, in the battle, he was known universally as Patton.

"Thorbjorn came to stop the lava flow. Patton tried to help Thorbjorn—to get his idea into practice. Patton made it work mechanically."

"He was the first one who really tried to fight back. Before he came, people were just saving furniture. He put iron in the windows. It was his idea. He was the first to start to do something. As the chaos got higher, his role was bigger. He was a great action man. He was in charge of the pumping."

"When the pumping began, I thought it was just crazy. I thought the same thing when he was nailing iron plates in front of windows. But he gave you orders. You kept your mouth shut and did what he told you. You did not so easily say no to him. He worked people out like I don't know what."

"He was used to being in command, and he could be very aggressive. He was disliked by many people. He was not a man to discuss things. He was a man to act."

"He ordered, on his own, the aluminum pipes from Keflavik. He always passed the system. You were supposed to go to the Civil Defense and they would call Reykjavik and a Minister would call Keflavik. He called a friend in Keflavik and the pipe was here in three hours. A hangar was under construction in Keflavik. He had the metal brought here for the windows. He never took no for an answer. If he thought he was doing the right thing, he did it. He took nothing from no one, and did what he thought was right. He was called Patton. He did not dislike it, I think. Not at all."

Thorbjorn, the archetypical scientist, was full of conceptual alternatives. Patton preferred ideas to report for duty one at a time. When Patton asked "Does it work?," Thorbjorn would say "It will do no harm."

These two met every day with, among others, Magnus Magnusson, the mayor of Vestmannaeyjar; Pall Zophoniasson, the town engineer; Thorleifur Einarsson, a geologist; and Gudmundur Karlsson, managing director of Fiskidjan, the island's largest fish factory. These "battle meetings," as they came to be called, were where strategies took form and decisions were made. More than once, they considered abandoning the island. After the wind shifted and the fire rain came, the command post made plans for a total evacuation, completing an exodus that had begun in the first hours of the eruption.

Thorleifur called Reykjavik and spoke with meteorologists. The wind would change, he was told. If that happened, the ash would go back to sea. The group decided to stay. As if in response, the firestorm intensified. Haroun Tazieff arrived, a French volcanologist of world renown, and joined the bivouac in the aquarium. Three million cubic yards of tephra were already on the town, and the amount of extruded lava was fifteen times that. The height of the growing volcano was nearly four hundred feet. As the firestorm continued, a bomb weighing twenty-two thousand pounds landed sixteen hundred feet from the crater.

On the sixth day, the wind veered southwest, and the ashfall returned to the ocean. The town seemed covered by deep black snow. Many houses were discernible only as dunes in the tephra. If they burned, they left kettle-shaped pits. (Eventually, some of the ash-covered houses filled with steam and were cooked until their frames came loose like the bones of stewing chickens.) When real snow fell, it made a black-and-white, silvery, surrealist world.

Tazieff gave his opinion that there was no future for Heimaey. He said another eruption was coming. He was concerned about emissions of hydrogen, and he warned that the island might, in fact, explode. Tazieff was without peer in the coining of cheerless predictions. The volcanologist Sigurdur Thorarinsson awarded the community at least a fifty-fifty chance, but said that the eruption could last three years. Other scientists offered the calming view that this seemed to be a type of eruption not uncommon in Hawaii, wherein the pressure drops in a short time and the flow is over. At the University of Iceland, the petrologist Sigurdur Steinthorsson and others studied the chemistry of the lava, with the encouraging result that most of what they found was a form of basalt called hawaiite. Some of the Icelandic scientists noted that while a big ashfall had been thundering on the roof of the aquarium Tazieff had awakened to see a catfish swim up to the glass and open its mouth in "a silent scream of terror." A North Atlantic catfish is not an easygoing ameliorative Mississippi channel cat. It is a hideous gray calico, no less malevolent than the cat in your kitchen. According to the Icelandic scientists, Tazieff's nightmarish encounter with the catfish provided the data for his lamentable prediction. He read in the face of the catfish that all nature had been shaken to its foundations.

A high-fashion model landed on the island, was photographed on the black ash against the fires of the volcano, packed up, and flew away.

On the eleventh day of the eruption, ash came down with winter rain in a mixture so dense that a person on a sidewalk could not see across the street. Bombs were falling through the ash and rain. Flowing lava, moving north and east, had by now extended the island about two thousand feet—land-filling completely some twenty-five fathoms of sea. On the thirteenth day, a fresh flow moved on a vector toward the south harbor wall—the breakwater nearer the volcano. Lava crept along the shoreline, its progress written in clouds of steam accompanied by flashes of lightning. The volcano's murmurings became out-and-out explosions, so violent they shook the town. People watching the volcano could actually see shock waves rumpling its mushroom cloud. Seconds later, they would feel the explosion. On the fourteenth day, the force of the eruption increased. The liquid sprays of lava reached heights around a thousand feet. Complete pitch-blackness enveloped the town at noon. Five million tons of lava were now in motion, closing the distance to the harbor wall. All ships were ordered away. The harbor was forsaken. On this day, February 5, known thereafter as Black Monday, virtually everyone on the island presumed that the battle was over—before it had really begun.

ALL OVER HEIMAEY are silhouettes of fishing boats on the sides of houses and commercial buildings. A large piece of outdoor sculpture suggests the unmerciful in the beneficent sea. Rock gardens have been made with volcanic bombs. Driftwood tree trunks—standing in the earth upside down, with their roots spreading out like branches above—are the nearest things in Vestmannaeyjar to native trees. With finely wrought signs, private homes are named and dated: Herdubreid 1925, Landakot 1916, Litlibaer 1904. Where an exterior wall lacks windows, odds are it has been used for a mural.

A mural on the upper half of a four-story guesthouse was

painted in 1977 by twelve-year-old schoolchildren, whose tenure on the scaffold must have thrilled their parents. The mural depicts the predawn hours of January 23, 1973, when the artists were seven or eight and their lives were abruptly altered. Families with suitcases are running on a pier toward waiting boats. Many cars are on the pier. In the background is Helgafell the Holy Mountain, an ancient volcano standing silent while a wall of fire fills the night nearby.

The eruption came without warning—that is, without sufficient portent to yield an interpretation resulting in alarm. Gudmundur Karlsson, visiting friends the previous evening, thought he felt vibrations. He asked his friends if they were having problems with their central heating. They looked at him oddly and said they felt nothing. In the hour after midnight, Magnus Magnusson thought he counted fifteen earthquakes. They were like touches of pallesthesia, nothing more: little shivers in the bones. Magnus turned off his light and went to sleep. This was Iceland, after all, where the earth is full of adjustments, like a settling stomach.

Two seismometers picked up the tremors. Both were on the mainland and within sixty miles of Heimaey. One was under the charge of a schoolteacher at Laugarvatn, east of the capital. The other was on the farm of Einar Einarsson, in the Myrdalur, on the southern coast. Einar had the recorder in his living room, and it is still there. The seismometer is in bedrock a third of a mile away. When an animal walks around at night, Einar can tell from the machine if it is a sheep, a cow, a horse. He can differentiate the local milk truck from other vehicles. When the Concorde flies over, it signs the seismometer. The machine can discern the heights of waves. The university installed it on Einar's farm about a year before the Heimaey eruption, its primary purpose being to sense the threats of Katla, an unusually dangerous volcano only fifteen miles away. Hekla is in the area as well—the stratovolcano that appears in early literature as one of the two mouths of Hell. Groans from dead sinners have been heard in the crater. But Hekla is out in the open, observable under the sky. The baleful Katla is covered with ice. It lies under Myrdalsjokull— a glacier field of two hundred and seventy square miles. When Katla erupts, as it has about twice a century, it creates a vast chamber of water under the ice. When the water reaches a critical volume, it lifts the ice cap, and one or two cubic miles bursts out as a violent flood—a blurt of water twenty times the discharge of the Amazon River. The outwash plains these floods have left behind are as desolate as the maria of the moon. A town, villages, and farms lie between Katla and the sea.

The day before the Heimaey eruption, Einar called Reykjavik and said he was seeing things he had not seen before. They were earthquake swarms and they continued for six hours. Reykjavik called Laugarvatn, and the schoolteacher said he saw them, too. The epicenters were about fifty-five miles from the schoolteacher, twenty-five miles from Einar. These two seismic stations were among the first of forty in a network planned to cover Iceland, and all are now in place. At that time, there were only the two functioning; and seismographs, for all their sensitivities, are not compasses. On a map in Reykjavik, the Civil Defense drew a circle around Laugarvatn with a radius of fifty-five miles and a circle around Einar's farm with a radius of twenty-five miles. The circles overlapped, and therefore intersected twice. One intersection was on the mainland, northeast of Katla and Hekla. The other was Heimaey.

Additional earthquakes came four hours before the eruption. It was later determined by analysis of seismograms that the magma had moved forty thousand feet upward in the earth in less than a day. Meanwhile, the circles intersected where they had before. If the third seismograph had been working, it would have completed the triangulation, locating the disturbance under Heimaey. Without it—and in the absence of complaints from anywhere—seismologists and the Civil Defense had been given a warning they could not understand. If the fissure had opened half a mile west of where it did, it would have split the town right up the middle and consumed it with lava.

Fifteen minutes before the eruption, the seismographs recorded the strongest jolt in the series thus far. It apparently marked the actual moment when the wedgelike force of the rising magma tore open the crustal surface, making the fissure, which—on and off the island—was more than two miles long. The woman whose central-heating system had been called into question happened to be looking out a window and saw the fissure open. Her home was that nearby. She and her sons got away, and have never returned to the island.

Pall Zophoniasson, in bed and half asleep, also felt that last precursive shiver. His wife, Asa Hermannsdottir, had told him somewhat earlier that she had felt a mild quake, and now this new and stronger one caused him to get up and put film in his camera. He remarked to Asa that perhaps there was a new Surtsey rising in the ocean somewhere near Heimaey. If so, he would photograph it from a plane. Then he looked east through a window and saw what he took to be a house burning—a red glow, a domestic accident. He called Magnus Magnusson, who lived on that side of town. Magnus said he believed there was an eruption somewhere east of Heimaey. The two got together and went to have a look. They saw the lava curtain rising, its fountains extending to the north and the south. Magnus took pictures. Pall took pictures. "We were like tourists," Pall remarked as he was telling me the story. They stayed half an hour.

The people generally did not panic. While aircraft were converging on Heimaey, and the National Emergency Operation Center was alerting ships at sea, the islanders' self-control seemed to grow in proportion to the menace of the spectacle before them. Quietly, they fled. Patients in the hospital and residents of the retirement home were flown to the mainland. Most people packed suitcases and walked or drove to the harbor. It was full of boats, thanks to the wild weather of the preceding day. Some people expected to be gone no more than a night. Others wondered if the island would explode, if a tsunami was about to wash over it, if ever they would see it again. In three hours, four thousand people left the island.

Dora Gudlaugsdottir, thirty-nine years old, got onto a net fisher called Home of the Gods. Roughly two hundred people were aboard; normally, the boat sailed with five. Her husband, who ran a small fish factory, was with her, as were their three younger children, including a four-month-old baby. The family had been asleep when the eruption started. For an hour and a half, they slept on. After 3 A.M., one of Dora's friends called. Dora told her friend to go back to sleep. She turned on a radio, however, and she and her husband soon realized the dimensions of the event. She made coffee.

"We didn't wake the children. We didn't want them to be afraid. Then we heard on the radio that we had to leave. We could see the curtain of fire. We saw it move to the north, toward the harbor. And we knew we had to go out of the harbor."

They packed one suitcase for the five of them (two other children were in Reykjavik, in school), and left for the quay.

"I saw no one crying. Everyone was still and calm. I think the Surtsey eruption prepared us for the next one. We knew it could happen. You began thinking, What if it happens here? The people were calm probably because they had thought about it and they knew what they had to do. Me? I was angry. I didn't want to go."

Dora now has a bookstore on Heimaey—actually, *the* bookstore on Heimaey—and is attractively bookish herself: a tall, slender woman, given to denim slacks and white pullovers, who wears reading glasses, has short tousled hair, and seems as generous as she is reflective. She told me that when she left her home she was mindful of Herculaneum and Pompeii.

"I had read about Pompeii, and I knew we had to leave. I knew there could be poisonous steam, as in Pompeii. Still, I was angry. Maybe I thought, I will never see this island again. I don't know. All the people had the same experience I did."

"You were angry at the eruption?"

"Of course. I couldn't be angry at anyone else, could I?"

When the boat cast off, she had gone below. Curiosity drew her up on deck. She took her daughter Ingibjorg, two years old, and held her hand as Home of the Gods, in a ruddy glare, passed between the tips of the breakwaters and under the sheer cliffs of the north side of the harbor. To reach the open ocean, the boat had to cross the fissure, and it did so in a chaos of sound and steam. Dora looked down into the water. She saw red lava there. In her words: "I'll never see that again, I hope. At first, I thought it was the glow from the eruption reflected on the water. I don't know if I saw it move. I must have been terrified. There came an explosion in the sea, and the boat shook very much." Salt water fell on their heads, and fine fragments of dark-brown glass. Ingibjorg was wearing a hat. Some glass stuck to the hat, and, in months that followed, no amount of scrubbing would take it away. Dora was bareheaded, and the glass went into her hair. "It didn't burn me. But when I came to Reykjavik I had to cut my hair. I couldn't comb it out."

Aircraft flying through the night toward the island saw something like a hundred boats, their lights brightly strung in a northwesterly line, heading toward the Iceland coast. On August 24, 79, Fleet Admiral Gaius Plinius Secundus, to be distinguished in history as Pliny the Elder, set about saving lives with quadriremes. Of course, there are differences between Vesuvius in 79 and Vestmannaeyjar in '73. There are similarities as well, however, which haunt the echoes of Dora's thoughts. Pompeii was buried under fifteen to twenty-five feet of tephra—roughly half a million cubic yards. In the course of the eruption in Vestmannaeyjar, three million cubic yards of tephra fell on the town, deep on the eastern side and thinning to the west. In Pompeii, as on Heimaey, many roofs fell in from the weight of the ash. Survivors of Pompeii soon returned, hunted for their houses, whose roofs were barely sticking out, and dug down to the doorways, so they could remove furniture. People of Heimaey did the same. Among those who stayed on the island to fight fires or pump water on the lava, some routinely shovelled pits in the tephra to get to the doorways of the houses where they slept. Pompeii, with four times as many inhabitants, was in area about half as large. Pompeii was a hundred and eighty-three acres. Pompeians were packed inside defensive walls, and they came and went through gates—the Herculaneum Gate, the Vesuvius Gate, the Water Gate. Vestmann islanders, while clustered, were immured by nothing but the sea, and they came and went through their famous harbor. As Vesuvius erupted, a wind was blowing across the mountain in the direction of Pompeii. That brought tephra down upon the city, first in very hot pieces the size of peas, grapes, and walnuts, and then in finer ash—but slowly enough to allow at least eighty per cent of the people to get away. If the wind had not blown from the west over Heimaey, carrying the initial tephra to sea—if the winter wind had, as usual, been moving in the opposite direction—people would have died, heaven knows what percentage of the population.

Pompeii was five miles from Vesuvius. The center of town on Heimaey was less than a mile from the crater of the nascent volcano. Vesuvius, beside the sea, is estimated to have been four thousand feet high before it blew off its summit. Eldfell—as the new, seven-hundred-foot volcano on Heimaey is now called—did not exist before the eruption. Like the people of Heimaey, the people of Pompeii felt some mild tremors beforehand but did not become alarmed. Soon after the Vesuvian eruption started, a large part of the periphery of the Bay of Naples was filled with panic-stricken crowds.

"There were those who, in their very fear of death, invoked it. Many lifted up their hands to the gods, but a great number believed there were no gods, and that this night was to be the world's last, eternal one."

Those sentences were written by Pliny the Younger, nephew of the admiral. To him the science of volcanology is indebted for its earliest preserved descriptions. As the huge column of erupting ash pooled against the stratosphere, it flattened out. Pliny wrote that it looked like a flat-topped umbrella pine. In the vocabulary of volcanology today, an explosively rising turbulent stream of magmatic fragments and magmatic gas is known as a Plinian eruption (Mt. St. Helens 1980, Heimaey 1973, Mayon 1968, Bezymianny 1955, Hibokhibok 1948, Krakatoa 1883). There is irony in the fact that the nephew has become in science the eponymous Pliny, for his uncle the admiral was only a reserve officer, as it were, and by calling was a naturalist. He can be likened vaguely to Rear Admiral Harry Hess, U.S.N.R., an earth scientist, whose "History of Ocean Basins" built the path to plate tectonics. Pliny the Elder, for his part, wrote the thirty-seven-book "Naturalis Historia." In 79, when he called for a vessel, he meant to head for the smoke and study the eruption. The idea of saving lives was an afterthought. Approaching Vesuvius over the water, he must have cut quite a figure, above the banks of sweeping oars. His nephew described him kindly as "somewhat corpulent." Today, he would be looked upon as somewhat sumo. The closer he got to shore, the more hot ash fell on him and his ships. When someone suggested that he turn back, he cried, "Fortune favors the brave!" The ships rowed on into the incandescent hail. Pliny remained—in his nephew's words—"so calm and cool that he noted all the changing shapes of the phenomenon and dictated his observations to his secretary." Harry Hess, skipper of an attack transport, dragged a Fathometer across the South Pacific and into the Battle of Lingayen Gulf, the invasion of Iwo Jima. Ashore, Pliny and his retinue went around with pillows on their heads, in much the way that people on Heimaey shielded themselves from tephra with pieces of corrugated iron. And suddenly he died. Pliny the admiral died, beside the bay, during the eruption. His nephew gave the cause as "some gross and noxious vapor"—gases of the sort that killed two thousand people in Pompeii. But no one perished with Pliny. His probable cause of death was myocardial excitement.

RAIN IN THE FINE ASH of Pompeii turned it into something like plaster, beautifully preserving the shapes of bodies.
Herculaneum, ten miles away and on Vesuvius's windward
flank, was buried by glowing avalanches of tephra and hot
gases (known in geology as nuées ardentes). Notably absent
was flowing lava. If there was any in the 79 eruption, it stopped
high on the mountain. After seventeen or eighteen centuries,
archeologists might arrive to scrape away indurated ash or mud,
but digs do not happen in basalt. The eruption on Heimaey
began in fields close to a farm called Kirkjubaer, in a setting
so beautiful that to think of it is painful for people who returned
to the island. The hay meadows and pastureland were a little
more than a hundred feet above the sea, to which they sloped
gently, everywhere presenting a view of nearby islands, of the
reassuring mainland, and, on bright days, of the not so af-
firmative Hekla. The farmer of Kirkjubaer sold milk in town.
On the night of the eruption, he shot his cows, left the farm,
and went to the mainland. He never came back.

Kirkjubær

There was prosperity in Vestmannaeyjar. Its fishermen, who were using rowboats as late as 1905, had since grown rich. After the people were evacuated to Reykjavik and began to wait out the uncertainties of the eruption, it was said of them that not the least of their rigors was the adjustment they had to make to the lower standard of living in the capital city. In their separation, they were not being supported in the manner to which they were accustomed.

On the other hand, while people left the island in boats, sheep went by air. Poultry was removed, too, and Iceland led the world in evacuated chickens. A horse went insane. A few days after the eruption began, seven hundred and fifty automobiles were taken to the mainland in a ship called Hekla. The town was full of empty houses, some with lights in windows, some with front doors open. A number of islanders had left Heimaey that night for the first time ever.

Just up the lane from Kirkjubaer, against the backdrop of the harbor cliffs, were Vilborgarstadir and the pond Vilpa, framed in transported stone. Vilborgarstadir was one of the first houses to be covered by the lava. Loftur Jonsson lived there, its last farmer. His wife, Agustina Thordardottir, was dead. He went to the mainland and never came back.

Vilborgarstaðir - Vilpa

Cooling the Lava

And not far away was Laufas, the home of the late sea captain Thorsteinn Jonsson, author of several books. His widow, Elinborg, was there when the eruption came, and his daughter Anna lived next door:

Laufás

There were multifamily houses that resembled hotels:

Á Landagötu - Stóru-Lönd

And a small hotel was much like a single-family house:

Hótel Berg (Tunga)

These buildings and many others that are under the deep basalt were preserved by these sketches, made in 1973, for the most part from memory, by Gudjon Olafsson, an office manager in one of the fish factories.

In the early seventeenth century, a pastor named Jon Thorsteinsson lived for fifteen years at Kirkjubaer, and after he died a stone monument was erected there in his memory. He was killed by pirates from Morocco who terrorized the island in 1627 and have come down in local history as "the Turks." The monument—a pylon six feet high—survives. It was removed from Kirkjubaer at the beginning of the eruption. A year or two later, it was returned; that is to say, in terms of the coordinates of latitude and longitude the pylon now stands where it stood. People looking up from the streets of the town see its silhouette against the sky. It is three hundred feet above its original base. Kirkjubaer lies under three hundred feet of basalt.

I went up there many times when I was on the island—for the contemplation and the stunning views. On one occasion, Magnus Magnusson went with me. With his blue tie, blue shirt, and gray tweed jacket, Magnus had a stout and Scottish look that I found inspiring. He spoke like a Scot as well, having practiced his verbal skills as a young merchant sailor in wartime ports like Aberdeen. He had a round and amiable face, dark bushy eyebrows, dark-brown eyes, a leathery bald crown, and a fringe of fleecy white hair like the steam coming off the rim of the volcano. Through rectangular gold-rimmed reading lenses he peered at the monument.

HER ER KIRKJUBAER UM 100 METRA
UNDIR HRAUNI.

THAR SAT SIRA JON THORSTEINSSON
FRA 1612, LIFLATINN

I TYRKJARANI 17 JULI 1627.
ANDLATSLORD HANS VORU:

"HERRA JESU, MEDTAK THU
MINN ANDA."

Magnus Magnusson was born in Vestmannaeyjar two hundred and ninety-five years after the pirate raid. With his schoolmates, he grew up in an ambience of unabated apprehension—an enduring fear that Turks in some form would return. Looking up from the inscription on the monument, he said, "I remember always being afraid of the invasion of the Turks, who killed thirty-four men and women and took more than two hundred as slaves. People fled into niches in the cliffs. The Turks shot them down with guns. Also, as children, we expected volcanic activity from Helgafell. We dreamt of it in the night and were a little bit afraid. In our minds, those were the two most dangerous things—the Turks and the mountain. Children grew up with these facts."

They grew up with other facts as well, presenting risks that were less remote. "Twice, this island—when the population was under three hundred—lost more than fifty men in one day," Magnus said. "In a hundred years, the island has lost five hundred lives into the ocean." Pointing northwest, toward a cliff known as Hetta (Skullcap), he said an edible root grew there called hvonn, and it had once been a staple of the island. Over the years, thirty-three people fell to their deaths while collecting hvonn. Off in the ocean to the north-northeast, he picked out a skerry where, in 1950, a boat heading in to Heimaey had lost its power and crashed on the rocks. "The boat had been stood up by the waves until it was nearly vertical," he said. "At that angle, that particular sort of engine tended to come apart in some vital place, and that seems to have happened. The crewmen who did not drown came up on the skerry. From Heimaey, they were visible, moving around. Many boats went out there, but to no avail. The men survived on the skerry for three days, but the wind force was sixteen and no one could save them."

In other words, the people of Vestmannaeyjar, who were emblematic of the people of Iceland, had lived since the year of settlement in the endless presence of disaster. They had obviously not been dissuaded by it, and had learned to subdue their residual fears. No matter how overwhelming a situation might seem to be, if there was any possibility of fighting back they had done so, and this seemed to have produced evolutionary effects, expressed in the battle against the lava, and much confirmed in the following story.

In the open ocean about three miles east of where we stood, Magnus said, a net fisher had gone down one night in

March of 1984. It was a steel boat, seventy-five feet long, with powerful winches, a crew of five. The bathygraphy there was very rough lava. The net spreaders got caught in it, and the captain used his winches to attempt to yank them free. The spreaders didn't budge. Meanwhile, the cables attached to them were slowly wound by the winches, and the fishing boat pulled itself stern first under the sea. It rolled over. Three survivors climbed up on the hull. The time was about 10 P.M. The ship's emergency raft was trapped and unreleasable. The air temperature was below freezing, the water not much above. No means remained available to create a distress signal. Lights of Heimaey were visible to the west. The three men considered their predicament for half an hour. Then—in their jeans, their wool sweaters—they slipped off the hull and began to swim. One died almost immediately. The two others—Gudlaugur Fridthorsson and Hjortur Rosmann Jonsson, the captain— swam side by side and kept talking. Birds, screaming in the darkness, swarmed around them. After a time, when Gudlaugur put something in the form of a question there was no reply.

Thereafter, he talked to the birds. In daylight, sailors who have fallen overboard have been found by shipmates who steered toward hovering birds. There was no hope of that in the dark of this winter midnight, but Gudlaugur—twenty-three years old—consciously struggled to keep his wits through dialogue with shrieking birds. He knew that confusion was among the first symptoms of hypothermia and if he became confused he would die. Always, he saw light on the island. He swam by preference on his back, but he thought that heat would be lost most readily through the nape, so he swam for long periods on his stomach. He swam about six hours—at least five times as long as anyone ever has in water that cold.

When he reached Heimaey, he found himself in a hostile

wave-battered niche in the new lava, up against a cliff. His purchase there being hopeless, he went back into the sea. He swam about half a mile south and, this time, climbed out on a broad flow of apalhraun, the sharpest and roughest texture of volcanic glass. Barefoot, he crossed it, and lost a good deal of blood. After he reached some grazing land, he saw a tub full of water for sheep, broke the ice with a fist, and drank. He came to a house about six in the morning, eight hours after his boat capsized. When doctors examined him, they could not find a pulse, and his temperature was too low to register on a medical thermometer. A year or so later, doctors at the London Hospital Medical College put him in a large tank—he is six feet four and weighs two hundred and seventy-five pounds—and hovered about with miscellaneous sensors while Gudlaugur reposed in water refrigerated to forty-one Fahrenheit degrees. After an hour, Gudlaugur was bored, and asked for television. The physiologists concluded that his sub-cutaneous fat closely resembled a seal's. In Iceland, where swimming is the national sport, Gudlaugur is not regarded as much of a swimmer. "He was fat," said Magnus Magnusson, as he finished the story. "He was no special swimmer."

Nor, by such standards, was Magnus a special diner when, in 1941, a bomb from a German airplane landed on a restaurant in Aberdeen. Magnus was at a table by an upstairs window. He could look down into the street. When the bomb hit the building, the front wall fell off. A waiter filled Magnus's water glass, and Magnus, who was hungry, did not lower his fork.

Magnus was nineteen. He sailed in North Atlantic convoys, and watched countless ships go down. He endured so many explosions that he lost a part of his hearing. This is not the Magnus Magnusson known as Mighty Magnus of the BBC, who has lived nearly all his life in Britain and translates Ice-

landic sagas. This is Magnus of Vestmannaeyjar, just one
among the seventy-six Magnus Magnussons listed in Iceland's
telephone directory. He has been described in Reykjavik as "a
quiet man of solid accomplishments," and these include some
years as a member of parliament from Southland, of which
Vestmannaeyjar is a part. At various times, he was Iceland's
Minister of Health and Social Security, its Minister of Communications. Democratically, he was retired from his duties
as M.P. and mayor. Not every mayor, of course, would know
what to do if his town split open and lava factioned the streets.
Magnus Magnusson knew what to do. There came a moment
in 1973 when Thorbjorn said to Magnus, "We cannot defend
the town and the harbor at the same time."

Magnus said instantly, "We must defend the harbor.
There is no use of any town if we don't have a harbor."

I reached into the ash at our feet, picked up a stone, and
juggled it in my hands like a hot potato. Magnus said, "Dig
two inches into the ash, you can bake hot-spring bread. Some
housewives do it. It takes four hours."

During the eruption, Magnus was among those who dug
down through tephra and in to his own front door. Although
the house was buried to the eaves, he continued to live there,
and shovelled his way in many times. (His wife and two children went to Reykjavik.) He has restored his house. It was not
one of the three hundred and fifty that went below the lava.
Looking down over Vestmannaeyjar with him from the pastor's
monument on this day of white clouds and light wind, I found
it difficult to imagine the rock beneath our feet at two thousand
degrees. Now it was known as the nyjahraun, the new lava,
and from its high advantage Magnus gestured into the topography of his home island and sketched it with its names, from
Hundradsmannahellir, a cave where a hundred men hid from

the Turks, to Blatindur the Blue Summit, to Lyngfellsdalur the Mountain Valley of the Lingonberries, to Horgseyri the Temple of the Old Gods. "Under the lava is the site of the first church in Iceland," Magnus said. "It was built before the acceptance of Christianity but in the same year." In Iceland, three dates that everybody knows are 874, when settlement first occurred, 930, when the parliament was founded, and 1000, when a great and prolonged debate resulted in the nation's accepting Christianity. During the deliberations, word reached the parliament that red-hot lava was advancing toward the farm of a prominent and influential Christian. To the heathens, this was a sign. One of them said, "The gods are angry."

The opposition replied, "Then what angered the gods when the lava we are standing on was flowing?"

Or so goes the story. Christian the country became, but in Iceland no amount of liturgy would ever clog the tube between religion and superstition. There was a long-standing omen in Vestmannaeyjar. One day, the son of a bishop would come to the island as a minister. Following his arrival, an eruption would take place. The Vestmann Islands parish had two clergymen. At the beginning of 1973, one of the posts was vacant. The church selected Karl Sigurbjornsson, and instructed him to report for duty on February 1. The eruption reported a week earlier. Karl Sigurbjornsson's father was Sigurbjorn Einarsson, Bishop of Iceland. The year 1973 would be a year of thirteen moons. This was not the sort of thing that generated faith—a year of thirteen moons. Also, there was an omen about the pond Vilpa, which was really a cistern—dry, ages empty, obsolete. The omen was that if Vilpa ever were to fill up, a disaster would soon follow. In the first days of January, 1973, the pond Vilpa filled up.

Cooling the Lava

AS WEEKS PASSED, the foremost question in everyone's mind was, When will the eruption stop? The topic was most often raised with Thorbjorn, whose patient answer was profoundly scientific. Thorbjorn said, "With each day that passes, it will be shorter by one day."

In the middle of February, when the lava was within a few hundred feet of irreversibly closing the harbor, the near-shore water was as warm as blood, and vapors were heavy above it. More pumps came. At Thorbjorn's request, Valdimar Jonsson, Professor of Thermal Fluids, came to the island as well. He brought his skepticism with him. Instantly, it went up in steam. "The water had the effect of putting up a solid rock wall," he has said. "I saw a fixed point that had been cooled while the lava streamed around that point. After a couple of days, I was convinced that we could do something about the lava, but we would have to pump at least ten times as much water."

Even the tide was endorsing Thorbjorn. At low tide, the lava by the water's edge advanced into the harbor approaches. At high tide, it remained almost stationary. One did not have to be a theoretical physicist to evolve a theory out of that.

Nevertheless, from Reykjavik there was as yet little support for the pumping—such a costly thing to do against odds that seemed to oscillate between hopeless and overwhelming. Bulldozers had utilized the tephra fall to sculpt long barriers, shaped like earthfill dams and positioned to divert the lava from the center of the community. In the end, they would prove ineffective, with lava moving both under and over them, but meanwhile they served as platforms from which the pumping crews could improve their trajectories.

As the harbor approaches narrowed, the pumping was

increased to thirteen hundred thousand gallons a day. Hoses set on tripods on the roofs of the cabs of bulldozers mimicked the appearance of flame-throwing tanks. The more steam, the more efficient the cooling. Hoses were moved when steam subsided. Thorbjorn came and went. The municipal airport in Reykjavik was only twenty-five minutes away. Patton was in for the duration, his face becoming bloodless on coffee, cigarettes, and condensed time. Patton was one of a number of people from the mainland who contracted what became known as the Vestmann Island disease. Valdimar Jonsson, who would have tested positive, described the symptoms: "If certain people stayed long enough, they found themselves unmissable. They could not leave the place. They felt somehow that nothing would run if they left." The syndrome pervaded the pumping crews, which, day after day, were made up of the same people. "They were just me and you and people on the street," Bjarni Sighvatsson remarked to me one day. "They were people from Iceland and people from Vestmannaeyjar. People left their families in Iceland and came here to work. A lot of people from Iceland came here to work."

The petrologist Sigurdur Steinthorsson was among those who monitored volcanic gas. He went around in an automobile, smoking fat cigars, while his wife, Helga Thorarinsdottir, was up on the roofs of houses shovelling tephra. "The roofs moved up and down," she has told me. "It was almost like standing on someone who was being sick. Heave. Rumble. And then the eruption would go off." The ash in the streets was dense enough to drive on, even where it was fifteen and twenty feet deep. There was method in the cigars: if they were snuffed out by a deadly emission of carbon dioxide, they would serve as monitors in themselves. The gas came over the lip of the crater, flowed downhill, and went through the town like a river. When humidity was high, the river was visible, with its eddies of blue haze. It suffocated cats. It stalled cars. People's heads were generally above it, but knees and elbows ached, breathing became labored, hearts pounded. In time, gas filled some buried houses and made them uninhabitable. Magnus Magnusson had to move. After the harbor reopened, a sailor who tried to loot a pharmacy—possibly hunting for drugs—died in a pool of carbon oxides.

Lynn Costello, aged twenty-two, arrived from the mainland on the first day of the eruption and remained for several weeks, working as a sound technician for the filmmaker Osvaldur Knudsen and his son Villi. Visiting the island was not a simple matter for someone with a camera. Journalists of all ilks were assembled in Reykjavik, flown to the island in an old Fokker, shown the fountains of lava, and flown back to the mainland. The Civil Defense intended that they *not* stay. So Lynn and Villi, who worked on the ground, had arranged their own transportation and secured from a friend a house in which to sleep and hide. They dug through tephra to the door. Osvaldur commuted from the mainland, shooting footage from the air. The work was inconvenienced by the time of year, there being only about three hours of natural light per day, but the pyroclastic spectacle made the darkness photogenic. Villi, who was twenty-eight then, has since become a sort of barrel-bodied St. Nick of a man with a russet beard that flows like lava. Osvaldur, who is no longer alive, made a full-time career of filming eruptions. Villi has continued in kind. It is possible in Iceland to have such a calling. With various chambered magmas Villi plays cat-and-mouse. He keeps in touch with pilots, and is ready at a moment's notice to be tightly circling an eruption cloud anywhere on a beat of three hundred miles. People arriving in Reykjavik can go to his "Volcano

Show," in an impromptu theatre in his house, to see what is new in the national volcanology. There have been twelve outbreaks since Heimaey. Villi is kept so active by his ever eruptive homeland that his production schedule lags behind the lava. He shows his audience rough cuts and narrates them in person, his voice somewhat viscous—magisterial and slow.

When I called on him in Reykjavik, he said of Heimaey, "It was like in a war. The sirens. The houses going up in flames. People from the U.N. giving statements. Fear of explosion from gas leaks. Bombs falling. It was a miracle that no one got hit."

"It sounded like Hell, with those huge lava bombs hitting the windows," said Lynn Costello, who is now Villi's wife. "It was terrifying to feel the force of those lava bombs coming down on the little town. You couldn't sleep. You were too scared to sleep. There was no place to take a bath. I didn't take a bath the whole time on the island." In the friend's house where they were hiding was a collection of films that could have turned Hugh Hefner into cryptocrystalline pumice. Lynn watched the collection bug-eyed, discovering wonders and anomalies of nature. Films like that were not just lying around in Phoenixville, Pennsylvania, where she grew up. They found an old storekeeper in the center of Heimaey who had spurned the evacuation. He supplied them with biscuits for a while and some odd bits of moldy bread until his supply was gone. Eventually, they became so hungry that they went to the mess hall where Patton's army ate, and they interviewed and filmed the cooks. Explaining themselves, they bandied the word "television." Finishing the story, Villi said, "After that, we could eat in Heimaey."

"First, it was a clean little town," Lynn said, referring to the early days when the fallout blew into the sea. "Then it got covered in black. Then it turned white in a calm, peaceful snowstorm. It was incredibly beautiful. When that lava wall was coming down the side of the island, they were out there with hoses—can you imagine? They actually stood there and hosed that lava wall."

"It looked a bit sort of far out, to say the least," said Villi. "The lava flow was so absolutely immense. But there was no alternative. They had to do something."

There were days of very cold weather. It was, after all, winter. "The heat made things more bearable," Lynn said. "When the wind was not blowing, it was nice to get near the volcano to get warm." One day, Villi went up toward the warm crater after positioning Lynn and her sound equipment in the cold near the base of the mountain. The air was biting. Lynn had not known Villi very long, and had serious doubts about staying in Iceland, as he wanted her to do. She had, in fact, pretty much decided to leave, and now, in the harrowing cold, she became certain that she wanted to go home. Plowing the snow with her feet to make large letters, she began to write I HATE YOU. She finished the "I" and was working on the "H" when she changed her mind. She felt ashamed. She knocked the arm and one leg off the "H" and made an "L." Her completed message said I LOVE YOU. The sentiment distracted Villi on the mountain. His vigilance relaxed, and a glowing bomb, much larger than he was, landed beside him with a reverberant thud.

At about that time, the growing volcano was seven hundred feet above sea level, its shape conical. Red molten fountains continued periodically to shoot high up from the crater, and lava spilled over the rim. On February 18, however, a small molten spring ominously developed near the bottom of the north side, and lava ran down from it like a small molten brook. Patton sent some pumpers to put water on it, but they had scarce begun when they were forced to flee. The whole north side of the volcano came loose. Thirty houses were destroyed in an incandescent avalanche. The north wall partially collapsed, then moved away from the rest of the mountain, and, with its great compressive weight, both complicated and accelerated the motion of the over-all flow.

Here began a sequence of events—as I count them, four—which occurred across a period of two months and basically influenced the future configuration of the island. They resulted in deliverance and destruction, on a large scale. They all involved human intervention, with effects that affected other effects and were ultimately so imponderable that no one could assign to people or to nature an unchallengeable ratio of triumph to defeat. The first took place at the south harbor wall.

This was the breakwater nearer the volcano—extending like a finger halfway across the harbor mouth. Continually hammered by wind-driven waves, the harbor wall was in need of repair, and was about to get it.

As the weight of the dislodged mountainside squeezed forth fresh pulses of underlying lava, the threat to the harbor redoubled. The pumping-and-cooling operation, still not taken seriously by the majority of Icelanders watching it on television, became desperate. Lava moving alongshore at the rate of a hundred feet a day was obviously going to reach the breakwater and then, likely, cross it and fill the harbor. This seemed so probable that the Civil Defense Council once again ordered all boats away. Big nozzles on tripods were lined up on the breakwater like cannons on a deck. When the lava had covered the final distance and was looming very near, water trajectories could reach over the front and more effectively soak the top.

Lava tends to advance somewhat lurchingly as hardened material tumbles from above and fresh spills break out below. Sometimes a few tons—black-hot or red-hot—will roll down the front, swiftly killing anything in their path. According to Villi Knudsen, "the safest way to approach lava is to have another person with you and he goes first." While the pumpers continued to throw everything they had up on top of the flow, Thorbjorn, Thorleifur, Patton & Company called for the help of a commercial ship called Sandey. Ordinarily, it worked in shallow water around Reykjavik, using a powerful pumping system to suck up seashells for use as construction material. (There is no limestone in Iceland.) Sandey's equipment could spew out six thousand gallons a minute. When Sandey entered the harbor, at considerable risk to its owners and its crew, it joined forces with a local boat called Lodsinn, which was rigged up with pumps, and together they quintupled the quantity of water being thrown on the lava, raising it to more than eleven and a half million gallons a day. In the words of Agust Erlingsson, who was one of the pumpers, "A big steam came out of that." From a large nozzle on Sandey's bow, water arched across the harbor wall and splattered on the lava, where large chunks in the flow were so hot they seemed to be melting like ice cream. For four more days, the future of the harbor remained doubtful as the ships and the land-based pumps engaged in their push-of-war while behind the lava front the oncoming flow buckled as it tested the resistance of the cooled margin. Then, wondrously, a condition of equipoise was achieved—a standoff, during which the lobe spread to left and right and filled in the sea bottom on the outside of, but next to, the breakwater. The volume of lava grew, and the front became higher, but held. The lava stopped.

"It stopped one foot short of going over the harbor wall," Magnus Magnusson told me.

"Not one stone went over it," Agust Erlingsson said. "I saw one stone roll on top of it and stop."

Gudmundur Karlsson said, "I saw three or four stones fall into the harbor, nothing more."

The breakwater, which had seen its last wave, was no longer a breakwater. It was now the lining of the inner edge of a new high headland, occupying one side of the entrance to the harbor.

LIKE AN ICEBERG that had calved off a glacier, the great bulk of the north side of the volcano remained afloat in a molten sea. It was a mountain in itself, and, moreover, it moved. It was landscape on the loose, an incongruous itinerant alp, its summit high above the lava plain, its heading north by northwest. The mobile mountain had a nine-acre base and a sharp peak. It weighed two million tons. People looking up from almost any street in town could see its silhouette filling the sky—today in one place, tomorrow in another. Someone named it Flakkarinn. And no one ever called it anything else. Flakkarinn the Wanderer.

The pressure wave that was created when Flakkarinn came off the volcano moved through the lava for a number of days and squeezed from the periphery new freshets of red rock. Some of this was in the lobe that stopped at the harbor wall. Flakkarinn, sliding downhill, also made bow waves in the molten lava through which it plowed. And as it went along it dug a kind of trough. Lava filled in behind it. Where Flakkarinn broke the crust of the earlier flow, fresh streams of molten material poured forth. People climbed up and rode on Flakkarinn. It shook as it travelled. In its first two weeks, it went half a mile.

If all of this had happened on a different vector, it might have been merely entertaining. But Flakkarinn was headed for the harbor. If one of its advance waves had nearly overtopped the harbor wall, what might be expected when Flakkarinn itself arrived at the same place? When the Wanderer reached the harbor, the harbor would become a hill.

A plan was developed to stop Flakkarinn. The dramatics at the harbor wall had amply demonstrated that pumped sea-water could affect both the motion and the final position of the right kind of lava. As Thorbjorn explained, "all this was possible only because the lava was thick, viscous, and moving slowly." (In what is now the United States' Pacific Northwest, an eruption once buried in three or four days an area the size of Iceland. As they say in Olympia, try watering that.) To mount an attempt to obstruct Flakkarinn, all available pumps were requested from the Americans in Keflavik, from the Civil Defense in Reykjavik—and transports arrived full of pumps. The strategy was straightforward: Select an area of the lava lying in Flakkarinn's path, and pump enough water onto it to get below the surface rind and increase in size and number the columnar cracks that characterize basalt as it cools. Then more seawater, saturating the cracks, would reach all the way to the impermeable molten center of the flow, solidifying an over-all mass sufficient to block Flakkarinn.

When Thorbjorn was reviewing these events with me, he said, "This ship Sandey, it had some steel pipes over half a metre in diameter that were very heavy and difficult to handle. After we got the bulldozers up on the lava and put the pipes there, the lava moved, and the pipes, of course, broke. There were some very courageous men who managed to keep the pipes intact, more or less, most of the time."

"It was iron pipe put together with nuts and bolts," Sigurdur Jonsson recalled. "The lava would move it a good distance overnight. While you were repairing it, if there was a big explosion in the volcano the air pressure came like a wave and could shake you."

Everything was moving, up on the nyjahraun. Where the crust was thick enough, the bulldozers operated even on moving lava. Where they put too deep a scratch in the surface, the red fluid welled up like blood.

"Water was pouring straight out of pipes onto lava," Sigurdur continued. "It was just like a waterfall. Thorleifur and Thorbjorn had figured out where Flakkarinn was going, and that point is what we cooled. We used Sandey against Flakkarinn. We also used the Vestmann Islands pumping ship. Also a raft in the sea with pumps. Also fire engines. Also pumps on shore. Everything you could pump with. With no harbor, there is no town. It is just a summer holiday."

Something close to thirty million gallons of seawater was poured into that one chosen area to prepare it to repel Flakkarinn. As test drillings eventually substantiated, the volume of basalt that was solidified for this purpose was about the size of Yankee Stadium. Toward it, Flakkarinn kept coming— toward the chilled lava and, beyond it, the harbor. The crash was rock-buckling, full tilt, head on, tectonic. After spinning halfway around, Flakkarinn broke into pieces that no longer moved.

THE DESTRUCTION OF Flakkarinn and the stand at the harbor wall were such convincing victories that the Icelandic government circulated a request for "movable high-pressure water guns"—for the most powerful pumps available anywhere. The United States Navy volunteered to look around. In a sense, the pumpers seemed to be ahead of nature by a two-to-nothing score, but it wasn't much of a sense. It was local, peripheral, lobar. The volcano had thus far added more than a million cubic yards to the bulk of the island, and a great deal more was to come.

Right after Flakkarinn disintegrated, a new and voluminous lava tongue broke out. Most of the magma that had come up in the eruption so far had flowed north and east, spilling into the open sea. The new tongue headed west-northwest, directly into town. It burrowed under and tossed apart the bulldozed barrier defenses. It advanced on a front a thousand feet wide, on its way to demolishing a fifth of the community. In a single day, it would cover as much as eighteen acres of streets and houses to a general thickness of sixty-five feet. It came to be known as the City Flow. When it broke loose, exactly two months after the beginning of the eruption, the cooling operation abruptly ceased, as pumping brigades were forced to take apart their pipelines and clear out. Equipment was lost. In the words of Gudmundur Karlsson, "We were retreating day and night. Sometimes we were retreating fast and sometimes we were retreating slow—but we were retreating all the time." It is widely said that in the battle meetings there was much confusion at this time. To confusion, though, there may not have been a rational alternative.

In one day, the City Flow took seventy houses. When the lava moved against them, some flared up like Ping-Pong balls ignited by a match, others were crushed like eggshells, others came off their foundations and were scuffed along through the neighborhood until they broke into pieces and were eaten. There were islanders on the mainland who flew to Heimaey after learning that their property was certain to be destroyed. Reverentially, they swept out and otherwise brightened the appearance of their homes, preparing them for the lava. In 1926, at Hoopuloa, on the Kona coast of Hawaii, a storekeeper swept his place clean before it went under lava.

Sigurdur Jonsson, remembering the City Flow, said, "We had to stop the pumps, take the pipeline apart, and throw it. Then we had to build the system again. We laid a new pipeline in town. Then we had to move it. We laid another. The City Flow went faster than others. You had to get everything away as quick as you could. It didn't matter how you did it. Pumps were on the lava. We had to pull them away. We worked two shifts. When we came on, if the guys on the other shift had lost ten houses, that was nothing. If they had lost three water hoses, they were no good."

It has been noted that a space the size of a baseball park was hardened with seawater in order to vanquish Flakkarinn the Wanderer. A great deal more than that had been altered by the pumping. Along the harbor approaches, the cooling had not just stiffened the new shoreline but extended the chilled margin some six hundred and fifty feet back toward the volcano. The water-hardened zone was nearly a hundred feet thick. It had shattered Flakkarinn, and clearly it could influence anything to follow. Among the natural patterns of lava flows, it was utterly anomalous. In a very certain sense, it was man-made. And it asked a question: By creating this occlusion, had the pumping actually caused the newly appearing lava to turn left and overrun the city? Phrased another

way, were the direction and devastation of the City Flow products of human intervention?

The answer lay somewhere on the spectrum between
probably and yes. In almost any contest, everything that happens affects everything that happens thereafter. A free throw
missed in the third quarter does not make the difference in a
one-point game. Even in something as primal as a volcanic
eruption, the component of human interference could apparently enter the narrative and, in complex and unpredictable
geometries, alter the shape of succeeding events. After the
human contribution passed a level higher than trifling, the
evolution of the new landscape could in no pure sense be
natural. The event had lost its status as a simple act of God.
In making war with nature, there was risk of loss in winning.

Water was not the only weapon considered for use on
Heimaey. There was talk of bombing the volcano. In one
unofficial scenario, a United States Navy ship would saturate
the mountain with high-explosive shells. In another, a pre-
cision bomber or a ballistic missile would lay a device in the
crater. In any case, the intent would be to blow out the mountain's east side, loosing a gush of lava into the open sea. Explosives were contemplated also as a way of breaking the
insulating crust over lava that might threaten the harbor or
the town, exposing the molten interior, and causing it to cool.
There was a possibility of unforeseen results, some of which
were investigated by Stirling Colgate, an American physicist
from the New Mexico Institute of Mining and Technology.
His calculations suggested that if bursting bombs or other artificial explosions were to cause a rapid exchange of heat between lava and seawater above it, a "runaway mixing process"
might occur in the three- or four-megaton range. This recalled
the day in 1883 when the island Krakatoa, in the Sunda Strait,

made a sound heard three thousand miles away and was spread throughout the world as so very much dust that it darkened sunsets for seven years.

Many other possibilities were studied for diverting, slowing, or stopping lava, but, with minor exceptions, they were hypothetical. There was no manual. The Battle of Heimaey was without precedent. In words of the United States Geological Survey, it was "the greatest effort ever attempted to control lava flows during the course of an eruption." Others had been scattered and abortive. In Sicily in 1669, when lava appeared on the south slopes of Etna, citizens of Catania went up the mountain with picks and axes to break the crust on one side of the flow and cause it to migrate toward the city of Paterno. Five hundred Paternesi came out shooting. This led to a royal decree: Henceforth, no one would molest lava. No one did, for three hundred and fourteen years. In 1983, as blazing lava poured down a shoulder of Etna on a vector that included a ski resort, a barrier was constructed to divert the flow in the direction of another ski resort. A barrier was waiting there as well. The construction effort, which had been prodigious, was directly inspired by the saga of Heimaey. The lava made a pair of christiania turns, missing both hotels.

In 1911, when the Hawaiian Volcano Research Association was founded, it adopted the motto "Ne plus haustae aut obrutae urbes": "No more swallowed-up or buried cities." Field artillery was contemplated as a means of living up to the commitment, also high explosives tied to the ends of sticks. When lava extends as a river, its upper reaches crust over, forming tubes that insulate the orange-hot liquid racing within. In 1935, Keystone bombers of the United States Army Air Corps bombed a lava tube on Mauna Loa. The source vent was high on the mountain at nearly nine thousand feet, but the advancing front, far below, was threatening the port of Hilo. The tube, blown apart, became effectively clogged with debris, and lava spilled to one side, yet the result was inconclusive, because the mountain stopped erupting. In April of 1942, with Hawaii blacked out and Japanese carriers at large on the ocean, Mauna Loa erupted again. The lava river flowed so fast it included standing waves. Once again, the direction of flow was toward Hilo. The eruption was a military secret. People spoke of it in whispers, lest it serve as a beacon to the Japanese. This time, the planes were B-18s—Bolo bombers. Their target was a natural levee hardened at the edge of an open flow. The bombs breached it. The lava spilled sideways. It ran alongside the original stream and rejoined it a short distance below.

It should be mentioned that these historic efforts to do battle with lava in Hawaii were few in number, futile in nature, and not conceived by Hawaiians. They were products to some extent of mainland-educated scientific minds and to a much greater extent of West Point–educated military minds. Hawaiians had lived with eruptions throughout Hawaiian history, and their primary way of dealing with the problem was through votive offerings. To this day, one sees strewn flowers at the edge of active craters, flowers in vases, offerings of tobacco, of food, and, most of all, of gin. The offerings are there to placate an irritable deity whose dark humors are expressed as earthquakes and whose rage takes form as molten fire. Pele. Among the forces of the earth, she is as powerful as any but her sister the sea. Hawaiians have been as susceptible to science as everyone else, but their passive acceptance of the errant moods of Pele remains intact even if their belief in her does not. In what is now approaching two centuries of recorded data, their Long Mountain—their Mauna Loa—has erupted, on the average, every three and a half years, and in that short

time has poured out four billion cubic yards of lava: enough
to pave Iceland. Mauna Loa has coated and recoated itself so
often that it has never had time to erode. Whatever the rain
may have taken away has quickly been replaced. The long
mountain is fifty miles long. Viewed from the edge of the
ocean, it is an astonishing trompe-l'oeil, because it is so
smoothly constructed that it appears in two dimensions and
presents a deceptive depth of field. It looks like a low friendly
hill, a singing dune, at worst a bald Scottish brae. You think,
I'll run up there and have a look around before lunch. The
long mountain is as high as the Alps. If it were dissected by
streams—given promontories and reentrants, serrated by canyons, invaded by shadows—it might look something like the
Alps. As is, it's just a massive shield, composed of chilled
magma, looking the way the Alps would look if a dentist could
repair them.

The big island of which Mauna Loa is a part consists of
five volcanoes, which appeared at different times and grew
close beside one another. Slightly higher and somewhat older
than Mauna Loa is Mauna Kea the White Mountain, thirteen
thousand seven hundred and ninety-six feet. The youngest
Hawaiian mountain that is visible above water is Kilauea, one
of the two or three most active volcanoes in the world. Kilauea
was in a state of continuous eruption for the entire nineteenth
century. Its present altitude is about four thousand feet. Its
summit crater, an irregular oval, is two miles one way and
two and a half the other. In addition to the flowers and the
food and tobacco and assorted measures of gin, there stands
on the rim a small compound of buildings that are in themselves a votive offering. They are the United States Geological
Survey's Hawaiian Volcano Observatory, the highest part of
which is a glassed-in room that closely resembles an airport

control tower, and from which—with respect to lava—no control of any sort is contemplated, attempted, or exerted. Thomas L. Wright, Scientist-in-Charge, who has seen Styx itself come out from under his office—seen an Amazon of new, flowing rock—says he can envision no rational alternative to "letting nature take its course." When I called on him at the observatory one spring day, he remarked that "the Hawaiian heritage is to be fatalistic," and went on to say, "They accept the renewal of land by volcanic eruption. There's no feasible way of dealing with large, continuing flows. Political issues have aligned the state and county in a fatalistic mode. The consequences of diverting lava from one place onto another would be unacceptable. To me the idea of putting explosives on the upper reaches of a volcano—of putting artificial things into a natural environment—is abhorrent."

Eleven thousand feet up the north slope of Mauna Loa the National Oceanic and Atmospheric Administration has housed an expensive collection of instruments in a few small buildings that could easily be lost in new rock. In 1986, NOAA built a large barrier above the compound, in the shape of the Greek letter lambda—λ—with legs a couple of thousand feet long, to divide any lava coming from above and make the scientific station a kipuka, an island in a river of lava. Barriers had been proposed in Hawaii before—in various configurations to protect Hilo—but this is the first time in the United States that large defenses have ever been set up to protect property from volcanic eruptions. The NOAA barriers are in a high remote place, and when lava does come against them it is not going to carom into someone's kitchen. Near Hilo, though, on the low slopes of Kilauea, that sort of thing could happen. As on Heimaey, lava deflected from one route could wipe out houses on another. And this is not Iceland, the home of the fair; this is the United States, the home of the lawyer. When Mauna Loa erupted in 1984, the state was asked if, in dire emergency, an attempt would be made to save Hilo. The answer was no. The Department of Land and Natural Resources regarded such a struggle as futile in the first place, and, moreover, could not imagine any way to deal with the legal consequences of lava diversion. The Hawaii County Civil Defense Agency's standing instruction to firefighters is to try to prevent lava from causing runaway fires but to make no attempt to stop or divert it. Hawaiian firemen spend a good deal of time tracking hot lava through people's yards. Say an ooze-out from the main flow is threatening to raze the house of Louis Pau. The Fire Department turns its hoses on the lava in an attempt to keep the house from catching fire. This is a situation of some nuance. You can't pour water on lava without in some way affecting the movement of other lava. In Pall Zophoniasson's words, "If you stop the lava here, it moves there." That would seem to be the message from the Battle of Heimaey.

The juxtaposition of Mauna Kea and Mauna Loa has created something that an explosives-maker would call a shaped charge. In the crease where the mountains touch are many rivers of frozen lava and one of flowing water. The crease is a natural conduit, a natural slot; and its lower end is Hilo. This is the second-largest city and second-largest port in two thousand miles. An article published in 1958 in *Pacific Science* described Hilo's situation in terms that soon echoed in Vestmannaeyjar: "The loss of Hilo harbor would be disastrous to the present economy of much of the island of Hawaii, for there is no other harbor in that part of the island capable of handling the cargo that moves through the port of Hilo." Hilo's beautiful harbor, like the one on Heimaey, consists entirely of cold lava.

The eruptions of 1852, 1855, 1942, and 1984 all started at the summit of Mauna Loa and all stopped within a dozen miles of Hilo. In 1855, the lava was five miles away, in 1984, only four. As Tom Wright said, "Hilo was looking right at it." Mauna Loa was capable of sending forth as much as twenty-five million cubic yards per hour. In 1881, a flow entered what is now a part of Hilo, and came within a mile of Hilo Bay. Diversion barriers were proposed, but were so controversial they were never built.

In 1973, the story of Vestmannaeyjar travelled around the Pacific Basin like a message writ in smoke. Inspired by the triumph in Iceland, the U.S. Army Corps of Engineers considered building a dam high up the crease between the huge volcanoes. The dam would create a reservoir, the source of water for the chilling of future lava. Confronted with these developments, Harry Kim, director of the Hawaii County Civil Defense Agency, pondered what to do. The next time Hilo was looking right at it, people would be mentioning Vestmannaeyjar. "I knew there would be questions posed with regard to this technique, questions I could not answer," Harry Kim told me. "So I invited someone to come from Iceland—anyone at all."

The person who went to Hawaii was Patton. When Auckland asked similar questions, Thorleifur Einarsson went to New Zealand. But Patton himself—Sveinn Eiriksson—flew to Hilo and conferred with Harry Kim. This was the first summit conference in the history of fighting lava. Unfortunately, there is no transcript. Sveinn Eiriksson has since died, of a heart attack, and Harry Kim is careful to say that Eiriksson—whom he found "enlightening"—cautioned him not to be influenced by the publicity that attended the Icelandic situation. Harry is an intense man, serious and dedicated, slim to the point of being weightless. Deep is his love of the terrain, before and after it is altered. When he said to me that Eiriksson had told him that nothing much had really been accomplished in Iceland, I was about as startled as I would have been to learn that Eiriksson's American namesake had let it drop that nothing much had been accomplished in the Battle of the Bulge. The dubieties of Harry Kim seemed to me to be as astute as they were pragmatic. When lava starts filling Hilo Harbor, someone is going to be expected to rally forces and stop it—stop it in the way that Patton did in Iceland—and someone's name is Harry Kim. "They did not stop the flow," he said. "They worked on a small lobe, not the main flow. If we were going to attempt anything like that, it would have to be cost-effective. I would not spend a million dollars to save five houses. And government has to be careful. We must not destroy one home to save another. Morally, we should not do it—even if you give me legal protection."

I asked him what he thought of the Corps of Engineers' proposed reservoir and dam.

"Ridiculous," he said. "A discredit to what was tried in Iceland."

It has been asserted in scientific circles that Mauna Loa will inevitably destroy Hilo. This has given Harry Kim a visceral antipathy toward some, if not all, volcanologists. He thinks their alarming predictions have needlessly frightened the people. He said, "The town was in a state of great agitation when '84 came." The numerals referred to a time when only three and a half million cubic yards of molten material was pouring from Mauna Loa per hour. "I think for the most part the panic was controlled," he went on. "But certain individuals create a mentality. The mentality says, 'We *must* do some-

thing.' If Madame Pele could not be stopped by the Pacific
Ocean, how dare mankind attempt to stop her with pumps?"

WHEN '84 CAME and a bright-orange river poured for three
weeks from Mauna Loa, an eruption of much greater duration
was simultaneously taking place on Kilauea. As on Mauna
Loa, the lava was emerging from a mountain flank, and not
from the summit crater. The Kilauea eruption, which began
in January of 1983, was to continue for years. It began half a
dozen miles from the Kalapana coast. Crossing roads, eating
through villages and subdivisions, consuming great acreages
of forest, it has several times entered the ocean. There have
been times of crisis when it was consuming a house every thirty
minutes, between quiet times when the lava stream, making
a ruddy borealis against the night sky, just crackles along slowly
on top of earlier flows. In a subdivision called Royal Gardens,
a family whose house was lost had the lava field surveyed and—
in the manner of the pastor's monument at Kirkjubaer—had
a new house built on the new rock, above the old site. Nearby,
I noticed a house that had been spared when the lava stream
divided, flowing past it on both sides. The owners were still
in residence. One result of the present flows from Kilauea has
been the wholesale devastation of numerous marijuaneries—
or homesteads, as they are called—established by agriculturists
from the mainland. Marijuana has been described as Hawaii's
foremost cash crop.

If, for the first time, you were to touch down in a jet at
Keahole Airport, west of Mauna Kea, you would not be altogether irrational if you were to look out the window and
decide that a distressingly imaginable navigational error had

brought you to Iceland. The terrain surrounding Keahole and the terrain surrounding Keflavik are much the same: rivers of lava frozen in time. Vegetation has scarcely reached them. At Keflavik, the jagged aa has a green blush of moss; at Keahole, the apalhraun is black, without a hint of life. The Hawaiian flow is younger. In 1801, it came down off Hualalai, a lesser volcano eight thousand feet high, and poured into the sea. There on the leeward side of the island, where rainfall is ten inches a year, the lava has remained essentially unchanged. Resorts have been sculpted in it like movie sets, landscaped with imported soils. The bunkers of designer golf courses are not concave and full of sand but—lovely in the green surrounding turf—solid black islands of undisturbed basalt. Use your wedge on that. Your hands sting for a year. If a long approach shot lands on one of those, it bounces to Tahiti.

Flows of many ages cover the Big Island like wax that has grown on a bottle. But the action at the moment is all in the southeast, in the evening shadow of Mauna Loa, where Kilauea is. At the Hawaiian Volcano Observatory, a waiver was presented to me early one morning by Christina Heliker, a volcanologist. In Vestmannaeyjar, I had walked a good bit on rock so young that its interior was still in liquid form, while steam, in a reversal of the artesian process, was running uphill within the volcano and dancing away from the rim. Evocative as that was—as palpably as it told its story—a recent flow is not a live eruption. Heimaey had been quiet for something under fifteen years. Kilauea since 1983 had not been quiet for fifteen minutes. The waiver was in the first person, the prose of a ghostwriter who had studied law. Over my signature I mentioned "unusual hazards to persons and property," and went on to say, "I freely choose to encounter such hazards on my own initiative, risk, and responsibility," and "I do hereby, for myself, my heirs, executors, and administrators, remise, release, and forever discharge the Government of the United States, its officers, and its agents from all claims, demands, actions or causes of action, on account of my death." After signing, I got into a fire-resistant flight suit, property of the same government, and a pair of what I hoped would prove to be heat-resistant boots, property of myself. Heliker's boots were new, and she thought them a little tight. She said, "Once they melt a little, maybe they'll fit."

We waited on a tarmac helipad with Ronald Hanatani, who is a native of Hawaii, and is therefore, like an Icelander, a birthright volcanologist. Heliker is a convinced volcanologist. In 1980, she was working somewhere in the Pacific Northwest on a minor conundrum in glaciology when Mt. St. Helens exploded. Young, mobile, adventurous, she just packed her gear and went there to ask what she could do. "Once you get red-rock fever, you are never the same again," she remarked now. As if that were a cue, the chopper came over the trees.

We flew to Camp 8—a makeshift platform tent resting on a kipuka in a newly frozen river of black-and-silver lava. It was not all hard, by any means. Below its surface were arterial tubes in which the unceasing flow was descending in the direction of the ocean, which was eight miles from the source vent, in the mountain's eastern rift zone. For a year or more, the magma coming up from the interior of Kilauea had been emerging in a lava lake—a new crater some hundreds of feet across. The outlet of the lava lake was the mouth of the main-trunk tube, and it was larger than the Lincoln Tunnel. The main tube and its distributaries traced sinuous patterns under the lava field, their integrity—here and again—imperfect. Bits of roof had fallen into the tubes. These skylights, as they are called, were windows into the inferno, which is exactly what was there. From a panoramic altitude, they were bright specklings of white-orange in the silvery-black surface of the stream. If the helicopter hovered close enough, you could see Cerberus. While we hovered above one skylight, Heliker said, "I wouldn't want to stand on the edge of that." And she added, "This whole field was forest a short time ago." The field was as wide as the Mississippi River in New Orleans. It narrowed with distance from the lava lake. Camp 8, a mile downstream, was literally the eighth base of its kind. In the course of the long eruption, Camp 1 had gone under the lava. So had Camps 2, 3, 4, 5, and 6. Camp 8 was on a small knob, covered with ohia trees and fern forest, in the middle of the flow. The knob was the summit of a cinder-and-spatter cone from an earlier eruptive story, now all but lost in the new lava, which stood a hundred and fifty feet higher than the previous surface. Another kipuka, not far away, was so new that the trees at its edges were still burning.

Hanatani shot a laser at established reflectors, to see if Kilauea was locally expanding—swelling with increased magma. Calling off numbers, he said that the first reflector was "roughly" 1,505.718 metres away. He and Heliker also used a theodolite to measure the angle between Camp 8 and the rim of the lava lake, and when they had finished that part of the work we pulled on gloves and walked up there. The gloves were to prevent cuts when you slipped and fell on this newly minted terrain. It was a surface known as shelly pahoehoe ("pah-hoy-hoy"). Very different from jagged aa, pahoehoe is the other general texture of newly solidified lava. They are the results of varying gas content and viscosity, and their appearance is consistent wherever lava flows. In a magnified way, the version we were walking on might have been peanut brittle. You weren't sure what would happen to your next step. The rock would break, and you would crash through—dropping six inches, a foot—and you put out your hands to brace yourself. The rock, being essentially glass, was very sharp. It was also hot, particularly where a tube lay below and molten lava was running there. We came to a skylight and inched toward it. Steam swirled above it but did not close off the view—of the racing orange currents of an incandescent river. By an order of magnitude, this was the most arresting sight I had ever seen in nature. The time spent gazing into it could not be measured.

Gradually, I began to think. Out of curiosity, I asked Christina if we were looking down into the near side of the tube or were standing over the middle and looking at the far side of the tube.

"The far side," she said.

If my legs still had knees in them, I was unaware of it.

A few minutes later, continuing uphill, I crashed through the pahoehoe and fell about a foot, hammering a tibia against an edge of glass. The blow was sharp enough to draw blood, I discovered later, and the resulting lump lasted two months, but I don't remember feeling pain. I attribute this less to terror than to pure excitement. We were nearing the lava lake.

On any given day, the elevation of the lake will vary, depending on the efficiency of the tube system that drains it. On this day, the crater walls—which were sheer—were forty feet high. The lake was just what the word says: liquid. It appeared to be covered with elephant hide. There was an overlaying scum produced by contact between the lava and the air. This gray surface was continuously moving, heaving, tearing. It would rip apart from one side of the lake to the other, sending forth a red-orange wave. Lava would dive beneath lava. Or one patch of it would slide past another. Or one acre would separate from another and move in an opposite direction—opening a rift of brilliant red. Perhaps not by simple coincidence, the lake's stiffened surface was imitating plate tectonics. It is sometimes said that the way plate motions function is the way a thick soup will behave in a saucepan as it simmers. Scums form on the surface, and, driven from below by the convecting soup, they move. They slide under one another, they slide past one another, they diverge—like the Eurasian, Persian, Pacific, North American, Arabian, Bismarck, Fiji, China, Philippine, Solomon, Turkish, Adriatic, African, Aegean, Australian, Antarctic, Caribbean, Cocos, Nazca, and Juan de Fuca plates. In the Archean Eon, before there was land, soft-shelled continents are thought to have slid about as scums, experiencing comparable tectonics. This is what was going on in the lava lake. The plates of elephant hide were passing one another on transform faults and coming together in triple junctions, like the plates that make the surface of the whole earth. In the lake, the plate boundaries blazed with volcanic fire. The air was acrid. After Mark Twain was at Kilauea, he said, correctly, "The smell of sulfur is strong, but not unpleasant to a sinner." The heat was so intense you could lean on it.

Behind us, about fifteen feet back from the edge, was a crack about ten inches wide that ran on as far as we could see and was parallel to the rim. When we first stepped across it, I had remembered Christina Heliker at Camp 8 expressing irritation with the air traffic that often congests around the lake. She said, "The fixed-wing aircraft are limited to five hundred feet. Helicopters are not supposed to fly low enough to cause a safety hazard. They fly ten feet off the ground. They're out of Hilo control, so they police themselves—and they don't. Three or four helicopter tours and a fixed-wing plane may all be going for the site at once. They're an accident waiting to happen. They hover near the lake. If they have news media on board, they're trying to get pictures of geologists near the lake. The place is deafening. Hearing is important to the geologists, and we have complained to the F.A.A. You'd want to hear the crack if your piece of the rim were about to cave into the lake. With so many aircraft, that could be difficult. Sometimes, the smell of jet fuel is as strong as at an airport. The rim is unstable. Frequently, you get collapses with no warning. If you're near the rim and you hear a cracking sound, you want to jump back. But you would not hear it if a helicopter was hovering overhead."

While we were there, we saw only one tourist helicopter. It hung around like a dragonfly. Then it left.

I wish not to exaggerate the danger, which, for the most part, was more apparent than real. It should be noted that in Hawaii in the twentieth century only one person has been killed by a volcanic eruption. Lava in Hawaii is rarely explosive. You can be quite near it and know what it will do. When the bombs were falling on Heimaey, the people below were just flush with luck. Workers up on the lava were known for a time as "the suicide squad." They gagged on sulfur and suffered burns, but no one died.

On a transceiver, Heliker called our helicopter. A few hundred feet from the lava lake, we stamped out a crude helipad in the shelly pahoehoe. We flew downslope and landed near the toe of the lava stream. On that day, it was four or five miles from the lake to the place where the lava came out of its tubes and flowed red and black in the open. The bright-red channels, perhaps twenty feet wide, were configured like a braided river. The color was ruddier and duller than it had been in the skylights near the source, the lava somewhat cooler, and viscous. The flow stood up ten or fifteen feet, and moved slowly—chiming, crackling—with an occasional heavy thud as a partly hardened blob tumbled to one side and broke. Bulges would develop, then break, laterally spilling red-hot lobes. If the skylights in the tubes had been portholes of the underworld, and the lake the earth primeval, this association with the flowing rock was a good deal more intimate, and, thereby, even more stunning. We were next to it—to hot red tongues like tidal pools, to a mass of jagged fins moving downhill like sails. Ron Hanatani, shielding his face with a canvas bag, dipped a hammer into the lava and pulled forth a dripping coagulation that he set on the ground. Blackening on the outside, red on the inside, it slowly cooled. I asked if I could borrow the bag and the hammer. Holding the bag in front of my face, I dipped the hammer into the stream, and discovered that the liquid lava had the elastic texture of egg-white chocolate mousse. When I set it down, none of it stuck to the hammer, the temperature of the steel being so different from the temperature of the lava. Meanwhile, the canvas bag had turned brown at the edges and was giving off smoke. My own temperature felt abnormally high, but perhaps not so much from the radiant heat as from a critical case of red-rock fever.

I waited for my new rock to cool, with intent to take it home, where rock samples lie around in abundance and typically have ages of five hundred million years, fifty million years, five million years. The age of this one was five minutes. While it grew older and colder, I walked around the toe of the lava stream, in part to have a look at the actual front and in part to indulge the utmost desire to pee. About fifteen feet from the toe, I stood at parade rest, and—pissa a hraunid—faced the crackling, crunching, pooling, spilling, steadily advancing lava. Until that moment, I had no idea how hot the surface was on which I was standing—the hard pahoehoe of an earlier flow. The water, falling, instantly turned to vapor, and barely had time to hiss.

ICELAND AND HAWAII in a sense are twins. They are geophysical hot spots, the two most productive in the world. According to the theory that at present describes them, they are places where heat of very deep origin has found a surface outlet. Radioactivity in the deep rock creates so much heat that it must find a way out, and some think it originates as deep as the core. Travelling upward in plumes of, say, two thousand miles, it eventually encounters the thin surface plates, and may help explain why they move. In any case, when the heat reaches the underside of a plate it punches through, and, as the plate moves, punches through again, and soon again, like the needle of a sewing machine penetrating moving cloth. The plates average sixty miles thick. Beneath them, the narrow plumes of rising heat essentially remain in place, like the navigational stars. While the Pacific lithosphere slides overhead, the Hawaiian heat source stays where it is, making islands. There are five thousand miles of Hawaiian islands, older and older to the northwest, reaching to the trench just east of Kamchatka. Almost all of them have long since had their brief time in the air, and have been returned by erosion and seafloor subsidence into the fathoms from which they arose. The northernmost part of this chain is named for Japanese emperors, and the oldest emperor is about to dive into the trench. While the Pacific Plate moves north-northwest, the fixed hot spot, as it manifests itself on the surface, appears to move south-southeast. In the state of Hawaii, the oldest major islands are Niihau and Kauai, which, as peaks of volcanoes, have been out in the air for five million years. Kauai's amazing beauty—its huge canyons, its fjordlike coastal valleys—derives from the fact that it has been inactive long enough to erode. As successive lavas and intruded magmas, the islands build up from the ocean's abyssal plains to surprising volumes and heights. Mauna Kea and Mauna Loa, from seafloor to summit, are by far the highest mountains on earth. Their base is a hundred miles wide, and they are thirty-three thousand feet high. Kilauea continues to build. The Pacific Plate continues to move. As Kilauea goes off the top of the plume, something new will rise. In fact, it is already rising, twenty miles offshore—a new Hawaiian island, twelve thousand feet high at this writing, and three thousand feet below the present level of the ocean.

All of that and then some is Iceland. Iceland is a hot spot, but not a simple classic, like Hawaii. Iceland's track is written in the crust of the North American Plate, but Iceland has come to be also Eurasian. Iceland is a hot spot at a spreading center. It is exactly astride the Mid-Atlantic Ridge, where the Atlantic broke open, and from which it has spread fifteen hundred miles in each direction. In the complexities of plate motions, spreading centers not only spread but drift as well. The Mid-Atlantic Ridge has drifted over the Icelandic thermal plume. Eastern Iceland is moving east with Europe. Western Iceland is moving west with North America. The rock at the two extremes of Iceland is Iceland's oldest rock (something over fourteen million years), and the two sides become younger and younger as you move toward the middle until you come to the Mid-Atlantic itself, the center of the country, where the rock is so young, so fluid, it has no age at all. Iceland is the geologic chocolate shop of this minor planet.

There are people who like to think that Iceland, which

is already of great strategic importance in the geopolitical confrontation between East and West, is rapidly becoming more so as it grows territorially into both hemispheres. Villi Knudsen warns the world, "You'd better watch out for the future. Eventually, we are going to take over." Meanwhile, one of Reykjavik's principal thoroughfares is called the Road to the Warm Spring. People wandering about after late parties lie down in warm brooks.

At Geysir, northeast of the capital, a large hole in the ground is the world's eponymous geyser. The old geyser is no longer forthcoming. It is full of water but not of action. It has literally been roped off. Close at hand is a young geyser. At five-to-seven-minute intervals, the young geyser swells tumescently, lets forth a series of heavy grunts, and into the sky shoots a plume of flying steam. Meanwhile, the old geyser just sits there—boiling. On special occasions, Icelanders know how to make the old geyser do its thing. They throw soap in it. It erupts.

From Surtsey's inception, scientists of many disciplines have carefully observed the island, taking advantage of any number of exceptional opportunities—for example, to observe the very beginnings of the process of forest succession, as seeds from who knows where invade and colonize freshly forming soil. For many months, they watched intently for the first green shoot to make its appearance in the black grit. This could offer insight so deep it might extend its illumination even as far as the Silurian world. At last, a leaf appeared, unfurled, and was followed by a vine—a tomato plant.

The Icelandic mainland, in length and breadth, is roughly the size of Pennsylvania, and the Mid-Atlantic spreading center runs down it in a swath about forty miles wide that in Pennsylvania would include Harrisburg and Gettysburg.

Heimaey, Surtsey, and the rest of Vestmannaeyjar are off to the south like Antietam and Manassas, and are also positioned on the line of maximum action. The Mid-Atlantic Ridge in Iceland is locally known as the volcanic zone. Eruptions are about as frequent there as harvests in an orchard. In the Krafla Caldera alone, toward the northern end of the zone, there may be, typically, ten eruptions in twelve years. The strato-volcanic Hekla, farther south, has erupted twice in a decade and twenty-three times since the founding of the parliament. Grimsvotn, under the Vatnajokull (the largest glacier in Europe), has erupted thirteen times in fifteen years. The Civil Defense Council meets "whenever there's a tremor," and is thus in constant session.

In 1783, a fissure opened in south-central Iceland that was fifteen miles long, starting an eruption that lasted eight months. Icelanders refer to it as the Fires of Skafta, which brought on the Hardship of the Blue Cloud. This—the Moduhardindi—was a haze that spread all over the Northern Hemisphere. The haze included large amounts of fluorine, and it affected sheep and people. There was hydrofluoric, hydrochloric, sulfuric rain. The sheep rotted alive. The people weakened. A third of the population died. For the others, there was famine. Meanwhile, the long fissure, strung with craters, was pouring out sixteen billion cubic yards of new terrain. One summer day, a big tongue approached the riverside church of the Reverend Jon Steingrimsson, who happened to be conducting a service. The worshippers were afraid. Reverend Jon closed the chapel door, blocking the view, and led the people in prayer—a moment that went into history as the Fire Mass. When he opened the door, the lava tongue had stopped. River water ponding against the lava had helped to stop it. In the winter of 1973, a pastor held a service in Vest-mannaeyjar that was widely regarded as a reinvocation of Jon Steingrimsson's Fire Mass. Afterward, the eruption continued. As it happened, the City Flow began that night.

In all, it covered fifty acres. It demolished and largely buried two hundred buildings. How much more would have been lost if there had been no pumping no one can say. In the background, the volcano was alternately sending up ash eruptions and lava fountains as the City Flow pushed through block after block. Nearing the harbor, it came up against the wall of Gudmundur Karlsson's fish factory. This was a major obstacle, the size of a city high school—full of computers, automatic doors, machines that rubbed skins off, machines that sliced fillets like boards in a sawmill. "The lava flow was always coming nearer and nearer and we had been retreating every day," Gudmundur told me. "It was just like being at war. Then the lava broke sixteen metres into this building. It came through windows on the second floor. We could have driven a jeep onto the top of the building. We were pumping water from the roof onto the lava. Without the pumping, this factory would have been completely broken down."

Today, a new street goes past the fish factory through a road cut blasted in the lava. If you follow the street to the east and then go along by the edge of the harbor, you come to Skansinn, a stone fort with walls ten feet thick. It was built beside the harbor in 1628, in the aftermath of the invasion of the pirates called Turks. The fort has stood symbolic for three and a half centuries, and the Turks have not returned. The City Flow reached Skansinn at the end of March. The lava climbed upward and over the walls. Against furious pumping, it advanced a few feet more. It reached the middle of the fort and moved no farther. The City Flow stopped—or was stopped—there.

THE LAVA, damming itself, built up so high during the cooling that it is now a hill beside the town. Roads of black ash traverse it, winding through the otherwise untraversable aa. When I was walking on those roads, steam came out of the road cuts. The new basalt looked exactly like basalt two hundred million years old. In places along the road where the views were good across the town and the harbor, red plank benches with yellow arms had been lovingly set about, turning the eruption into a preserve, as we have done in Northern California, at Lava Beds and Lassen.

The view over the community is of red, green, blue, beige, yellow bright rooftops, walls of oyster and cream. Silver. Turquoise. Copper. Butter. It's a trig and colorful, prosperous, handsome town. There is a house in three shades of green that closely resembles the geologic map of Nebraska. Its appearance is not singular in Vestmannaeyjar.

Streets that once ran east out of town are culs-de-sac now in the lava. Captain Street, so called because many skippers own houses there, once had the highest views in town. It now dead-ends in lava. There is new construction in many places beside the halted flow. Up the street are houses with deep tephra still on top of them, windows vacant, walls askew. Some are just hunks of plaster-covered concrete attached to reinforcing rods twisted like matted hair. A horizontal chimney protrudes from a new black slope. Dora Gudlaugsdottir, in her bookstore, showed me a picture of preeruption Heimaey, saying that it was very beautiful then, with the clear tone that an aesthetic disaster cannot be stopped with spray.

Cooling the Lava

When the eruption was less than two weeks old, the Icelandic parliament increased the national sales tax, and in other ways created a repository known as the Catastrophe Fund. It paid Patton's army. It helped offset damages. The surtax has never been rescinded, and has evolved into an insurance system against natural disasters. In 79, the Emperor Titus provided emergency funds with his own money as well as money of the state, and the Roman Senate declared that the assets of extinct families of Pompeii and Herculaneum were to be used as a catastrophe fund.

From the perspective of the streets of town, the skyline of the lava is high and prominent, inclining upward toward the new volcano. On a fine blue morning after night rain, the ridge was steaming like a bread pudding, and tour buses were on it, as they were in every weather, slowly traversing the tephra roads. "There are now three tour buses," Magnus Magnusson said. "Half the tourists come from Iceland." He went on to say, "The north end of the lava is so high because of the cooling. The molten lava built up against the chilled lava. You can never see in Iceland so steep a lava front."

One house caught in the edge of the lava is to be saved indefinitely as a reminder of the flow's destruction. A large empty window stares out of the rock as if it were the mouth of a cave. Just inside is a steam radiator, all but enveloped in lava. Above the window is a red gable. The rest of the roof is crushed into the living room. The concrete walls are cracked like shattered glass. Magnus said, "Now the community owns that house, paid for by the Catastrophe Fund. It was a new and very solid house, one of the strongest on the island, a big house. It was decided to let it stay."

When I asked Magnus who owns the new lava, which increased the size of the island by twenty per cent, he said that it belongs to the Vestmannaeyjar community, because the community owns the whole island. People own their houses, but they lease their lots. Reflecting on the three hundred and fifty properties that were buried by the eruption, I remarked to Magnus that sorting out such a thing in the United States would take three hundred and fifty years. Someone had pitched a bright-blue tent high up on the flow, beside a piece of Flakkarinn, as shelter from the wind.

Magnus Magnusson was defeated as mayor after the war against the lava. The petrologist Sigurdur Steinthorsson has compared the fate of Magnus to the fate of Winston Churchill. On Heimaey, there were changes in the constituency. Pall Zophoniasson, who was mayor when I was on the island, pointed out that a third of the evacuees had not come back. "We have about twelve hundred new people," he said. "About half of those have no connection to Heimaey. Old-timers say the place is different now, but they don't agree how." One difference was that these comments, ex officio, were being made in a large hot tub beside Vestmannaeyjar's new swimming pool, where the delectable warmth at differing levels in hot tubs and the pool was coming from heat exchangers up on the new lava. The water, piped from the mainland, was melted glacier ice. The mayor—every inch a statesman, even in a tub—was a lean dark-bearded man who would not look amiss on a five-dollar bill.

Before the eruption, the population was fifty-three hundred. Gradually, it had reapproached that figure, but it was still six hundred shy. "The first years were not that pleasant," Gudmundur Karlsson told me. "It was like living out in the desert." When the wind blew hard, flying pumice made such a din against people's windowpanes that they had to shout to be heard across a table. Pall Zophoniasson told a reporter from *Iceland Review*, "The pumice-grit gets through every crevice, even into people's souls, if I may say so." Many who lost houses stayed on the mainland. Even the skipper of a fishing boat, who had left the island with his wife and four children, never came back. Just as the lava had covered fields, more green fields were quickly covered with new streets and new houses in an effort to bring back the people. "We miss that land quite a lot," Sigurdur Jonsson said. "It was a double destruction. It was necessary—to build quick, to try to draw them back. But many did not come. Many who did come soon went back to the mainland. They'd got the feeling of living in Iceland."

There was some incidence of what doctors had come to call "the Surtsey neurosis"—a writhing sleeplessness and acute anxiety that had made its clinical debut not long after the sea boiled and Surtsey emerged, erupting. "We thought it would never leave the people," a doctor told me. "But it has. People recovered very fast."

Not all marriages recovered very fast. "If your marriage has a crack in it, and then you have an eruption, everything comes out in the open," someone explained to me. "It's like in John Steinbeck, in 'Busspå Villovägar,' when the bus gets stuck out in the country." In a number of instances, one partner did not wish to return, the other did. The reluctant party was often a mainlander who had married a person in Vestmannaeyjar, and the prenuptial understanding did not include an eruption. The words *busspå villovägar* are Swedish. The name Surtsey derives from Surtur the Fire God. The people of Heimaey smoke like volcanoes.

Islanders who did not go back to Heimaey created opportunities for others. Vestmannaeyjar is often described in Iceland as a place of "mysterious attraction," invested with "an aura of adventure." One of the odder effects of the eruption is that it attracted people to live on the island, where the mystery was subsidized with cold cash—with jobs to be had in fishing, jobs in new construction. "I am not afraid," a newcomer remarked to me. "The south of Iceland, it can all blow up. Yes. It's like living in San Francisco."

Magnus Magnusson commented that when the newcomers refer to the mainland they do not call it Iceland. They use the word *fastalandid*, which means steady land or sitting land. "They don't have the old sense of Iceland and the Vestmann Islands," he said. "You'd have to be here from birth."

"We in many ways regret this change," Gudmundur Karlsson said. "The outlooks of the islands have changed completely. Many of our friends and relatives didn't come back. New people came. It was like changing a crew on a ship."

After the eruption, Dora Gudlaugsdottir and her family went to Eyrarbakki, because it was difficult to get an apartment in Reykjavik large enough to accommodate five children. Their furniture followed them to the mainland. They were in Eyrarbakki until May and then went to Selfoss. Having left Heimaey for what they thought would be a few days, they returned after half a year. "I thought, It is like a ghost town," she told me. "I knew what I was going to see. I knew we could clean it up and make a home again. I am very stubborn. I didn't like the lava. It was all black. I thought it was very ugly. Now I can see it has a beauty, too."

Gudmundur Jonasson, who was fourteen at the time of the eruption, said that he would regret all his life "the spoiling of the beauty of the island."

"The life has changed," Sigurgeir Jonasson said. "We lost very good people—islanders. When we walk on the street, we do not see the old people. The new people cannot think like we do about the mountains, the sea, the birds. Maybe after many, many years. The east part of the island was the greenest part. We had the farms. Now we have the lava. We don't like the lava, we people who were on the island before. We saw all the sea. But we got a much better harbor, ash for paving streets, and warmth to warm up the town."

On the summit of Helgafell the Holy Mountain, on the highest part of the crater rim, is a circular bronze tablet equipped with a swinging arrow. Helgafell, in whose shadow Magnus Magnusson and Dora Gudlaugsdottir and Sigurdur Jonsson and Gudmundur Karlsson and the other islanders grew up, has green, friendly slopes that become cindery only near the top, where the view is comprehensive. One clear day, after climbing up there, I slowly moved the arrow, stopping where the tablet informed me that the arrow was pointing at Hekla. I looked up, over water and far into the mainland, where the stratovolcano stood white under fast-passing clouds. To the west, the Reykjanes Peninsula was visible, sixty miles away. Turning the arrow southwest, I stopped it on "Hellisey," "Geldungur," "Geirfuglasker," "Brandur"—volcanoes of the Mid-Atlantic Ridge, each freestanding as one of the Vestmann Islands. When the arrow was pointing at Brandur, it was also pointing at a much larger island eight miles farther away—of which there was no mention on the tablet. Swung to the northeast, the arrow pointed straight into the unvegetated slopes of the new volcano—Eldfell the Mountain of Fire. But the tablet mentioned only Bjarnarey, an offshore islet that was behind the volcano and blocked from view. The tablet had been placed on Helgafell in 1952. The scene obviously required a nimble cartographer, working in a medium less eternal than bronze.

Where Heimaey used to have one conical volcano dominating its central landscape, it now has the two. Although they erupted five thousand years apart, they came up almost in the same place and stand side by side, at virtually the same height, their craters separated by less than half a mile. Helgafell appears to have cloned itself. The before-and-after aspect is exploited on picture postcards. In the hospital one day, some years after the eruption, an old man sat up crying, "Where am I? Where am I? Where am I?" He had been bedridden since before the eruption, evacuated to Reykjavik, and eventually brought back. Attendants hurried to soothe him. One of them said, "You are in Vestmannaeyjar, where you have lived all your life. Don't worry. Everything is all right." The new volcano was framed in the window beside his bed. "I am not in the Vestmann Islands," he said. "If I am in the Vestmann Islands, what is that mountain doing there?" In the Radhus, in the chamber of the town council, is a large painting of a man in rubber boots, a knitted cap, walking on volcanic ash on Birkihlid Street among half-buried houses. The two mountains are in the background. On another wall of the room is a painting that dates to the eighteenth century, showing a schooner in the harbor, a couple of dozen buildings near the shore, and one mountain.

I went up the new volcano one weathery afternoon and, from the rim of its broken crater, looked down on the new lava. From the chimneys of the heat exchangers, steam was streaming downwind. The vestiges of Flakkarinn stood up like avalanche debris. To the south, great heaps of tephra, like mounds of bituminous coal, marched across the fields to the ocean in a straight line, marking the fissure that split the island. The heat exchangers were first put in wells, but they corroded, even if they were made of stainless steel. Drill bits explored the bottom of the flow, passing en route through the core of molten lava—a feat that had been accomplished only once before, by a drilling crew in Hawaii. The engineers wanted to find out if trapped seawater had become a reservoir of super-heated steam. It had not. In the end, the method by which the community turned the lava into a central-heating plant was much the same as the method by which they had fought it. Where the new flow was thickest—more than three hundred feet—they spread hoses over an area of about an acre and dropped water through cracks to the molten horizon. At the time, that was sixty feet down. Since the molten horizon was impermeable, the water could not escape. It was captured as steam and circulated through the heat exchangers, saving the community two million dollars a year. The volume of the molten core was diminishing, though, and would soon become an igneous mush, ending the bonanza.

Hearing a sound like a chain saw's, I looked down to see a motorcycle climbing the new volcano. The rider wiped out as he approached the crater rim. The motorcycle, pinwheel-ing, spun downhill in a sputtering swirl of tephra. Ash over teakettle, the rider tumbled, too. At last arresting his fall, he got up, ran after his loose machine, trapped it, then resumed the steep climb. Reaching the top, he zipped around the nar-row rim. He stopped for a moment, contemplating. Then he plunged hundreds of feet down the dark-red slope inside the crater all the way to the bottom, under control. Motorcycles seem to have more prestige in Iceland than in, say, Greenwich, Connecticut, or Bryn Mawr, Pennsylvania. All over the new volcano, white-painted automobile tires mark the established tracks and competitive routes of what were described to me as "the largest, most powerful, most beautiful motorcycles in Iceland."

Where the lava went eastward into the sea, some of the

new cliffs are a hundred feet high. Waves occasionally break over them. After a Belgian trawler crashed there in 1981, waves threw pieces of the trawler up on top of the cliffs. Rocks weighing half a ton have been tossed up there, too. It was at the base of one such cliff that Gudlaugur Fridthorsson first landed on the night of his long swim. In two places, bays with rocky beaches have indented this new coastline. One is called Baetur the Beach of the Insurance Claim. The other is Vidlagafjara the Beach of the Catastrophe Fund. The rocks, whether they are the size of potatoes, grapefruit, watermelons, water beds, or motor buses, are all rounded. Most are so cavitated by former gas bubbles that they suggest black coral, or models of the human brain. It was discovered on Surtsey and confirmed on Heimaey that such boulders do not require thousands of years to assume their shape but can be rounded by a single storm. During the eruption, a lighthouse on the eastern shore went below the lava. A new lighthouse stands above the Beach of the Insurance Claim.

BEFORE THE ERUPTION, the harbor entrance was half a mile wide. The lava narrowed that to five hundred feet—a great improvement, doing away with the danger from eastern winds and waves. With Magnus, I watched a dredge taking tephra out of the harbor. A heavy black slurry poured from a large pipe. Magnus said the dredging had been continuous for what was nearing fifteen years. At quayside—in a thicket of cranes, booms, and stays—were trawlers and other fishing boats, four abeam. A few were small and double-ended, no longer than canoes. Top-heavy with radar, they were obviously undaunted by the North Atlantic.

Walking past the boats one day, I saw a young seaman,

who was standing on the quay, reach over the water with a paint roller that failed to touch the hull of an eighty-ton trawler.

"Almost but not quite," I said.

He set down the roller and asked where I was from.

In his running shoes, bluejeans, and corduroy jacket—under a tumble of light brown hair—he could have been from anywhere. His name was Ingi Thor Sigurjonsson, and he was from Hafnarfjordur, not Vestmannaeyjar, but he clearly had developed pride in his surroundings. "It's a totally natural harbor," he informed me. "The lava really made it good. It is said to be the best natural harbor in the world. In the North Atlantic, a harbor means everything. This harbor has the record for fish on land—more fish per fisherman than anywhere else. Boats come in so heavily loaded that only their bows and sterns are out of water. There have been times when the freezing plants couldn't handle the fish. To store fish, they had to use the streets and fields."

"When was that?" I asked him.

"Like ten years ago," he said.

Ingi went to high school in Texas. His mother took him there after she married an American geologist. Now, as a netman on the trawler, he was making as much as six thousand dollars a month. Auriferous is the aura of Vestmannaeyjar. As much as a third of the value of a haul is allotted to the crew of a fishing vessel and distributed in prorated shares. Ingi's trawler was the oldest steel fishing boat in Iceland, but its fish finder was among the newest, and could easily tell a pollack from a haddock and distinguish both from a cod.

At noon one day, I bought a prefabricated sandwich at an Esso station, and, since I couldn't read the label, asked what was in it.

The young attendant said, "I don't know how you call it in English. I think you say lobster."

I thanked her, and walked out to Eidi, an isthmus lying between two of the sea stacks that form the north side of the harbor. While I sat on a boulder looking over the water and consumed the state flower of Maine, a girl and boy, maybe six and eight, rode up on tricolored bikes. They were wearing Nike warmup suits. On the back of one bike was a cardboard carton, from which the boy removed a young puffin. He carried it down toward the edge of the sea. Holding it with both hands, he made four false heaves, and the chick flapped its wings, getting ready for flight. The fifth heave was real. The chick flew over the water, set down, and began at once to swim with the competence of a loon. Within ten seconds, it had dived for a fish. Now, in a Volvo, came a man and two boys. They released four puffins in the same manner. The man heaved them high. At first, they wheeled giddily in the air, as if to go back, but then they turned and sailed across the water—six, seven, eight hundred feet—before touching down. Immediately, they dove for fish. The children with the bikes took another puffin from their box. The little girl heaved this one, and it landed on the rocks. She helped it into the sea. A maroon Dodge Aries arrived, driven by a woman in a yellow running suit, with four children, two boxes, sixteen puffins. Baby birds—all prepared for flight in the same manner and then heaved into the air—were hitting the water like raindrops. No puffin took more than ten seconds before making its first dive—with one exception. The bird sat there indefinitely. Fishing was not his thing. More parents, more children, more puffins came to Eidi: a man and a child in a Jaguaresque Datsun 260Z, a man and five children in another Volvo, a party of two adults and ten children in a Citroën and a Mitsubishi van, a woman and a very small child in a Saab GL.

In spectacular belays from cliffs, people of Heimaey gather the eggs of countless varieties of seabirds. The baby puffins come to the town. When they are ready to leave the cliffs where they are born, the chicks at nighttime go for the light. The children have collected twelve hundred and fifty in a day, more than sixteen thousand in a season. Touching is the scene when the chicks are released, but as adults they are caught and eaten. A chef in town is a master of hot smoked puffin—a taste that marries corned beef to kippered herring. Burgundy flesh is in the puffed-up chest of a puffin. It is served with new potatoes and sweet thin slices of pickle.

Under the lava is Olafshus, home of the late Erlendur Jonsson, master catcher of puffins. Olafshus had territorial rights to Alfsey, which is two miles from Heimaey. He and a crew of three or four sat in high niches in the dizzying cliffs of Alfsey, and with long-handled nets caught puffins in flight. In a four-week season, Erlendur would be content if he returned to Olafshus, and his wife, Olafia, with twelve thousand puffins. Olafshus went very quickly under the lava. The puffin is among the nation's emblematic birds. With its bright-white chest, its orange webbed feet, and its big orange scimitar bill, it could be an iced toucan.

The cost of cooling the lava was one and a half million dollars. The lava brought more than thirty million dollars' worth of heating to the town, and harbor improvements worth a great deal more. Heimaey had lacked paving material. With the new tephra, the airport runways were widened and lengthened, and streets were freshly surfaced all over town. It is said on the mainland that people of Vestmannaeyjar are sensitive about the infrabooty they got from the volcano—about their favored-nation status with regard to the Catastrophe Fund.

They are not receptive to banter on this topic. As a mainlander noted, "you say that once, and not again."

The American pumps arrived in the tenth week of the eruption. To shoot water thirty feet into the air, almost any strong pump will do. If you want five hundred vertical feet, you need unusually specialized ordnance, and these machines had come out of a military warehouse at two and a half tons the unit, accompanied by papers calling them invasion pumps. There were nineteen. They were of the generation that was developed to move gasoline from offshore ships to Omaha Beach. Their aggregate maximum output was thirteen thousand gallons a minute.

In a two-stage procedure, they were supported by suction pumps, drawing water from the harbor they were trying to save. Altogether, about forty pumps were lined up in ranks on a broad quay, which came to resemble a natural-gas field or an oil refinery, with large pipes in parallel clusters and valves controlled by wagon wheels. The water went off the quay, through the town, and up on top of the lava, where steel pipes and aluminum pipes had been broken so often by the restive basalt that Valdimar Jonsson, who by now was in charge of the design engineering, risked the flammability of flexible plastic. Protected from within by the cold water, the plastic did not melt. Altogether, he soon had seven miles of working pipe, most of it up on the flow, discharging water directly onto the lava through diameters the size of dinner plates. The system pumped as much as twenty-three million gallons a day. The water ran downhill in simmering brooks.

The invasion pumps cooled the lava for a hundred and three days. On April 22, Easter Sunday—about three weeks after they began—a new lava current emerged from the crater that was thinner and faster, and therefore less predictable, than any that had come before. It bubbled. It surged. It seemed to come forth at a rolling boil. It ran east, fortunately, and veered south. Swiftly, it entered the sea. It was truly a red river—broad, fast, braided—moving like a big arctic stream. It flowed for more than two months. The greater part of it continued to flow east, building a new platform of about a hundred acres into and far above the sea. Some of it went north, in the direction of the harbor and the town, and ran into a very large reception of pumped salt water.

Valdimar said, "At each place we cooled, it was just like putting a nail through the lava. After three days, we would move the pipe thirty metres, and drive another nail. If more lava came, we were ready. We had put up our piping system to within two hundred metres of the crater. We were ready to take on new bursts."

When Valdimar was called in to set up the high-pressure pumps, he had been Professor of Thermal Fluids in the Engineering Department of the University of Iceland for less than six months. He had taught at Imperial College in London. He had taught at Penn State. His Ph.D. is from the University of Minnesota. He grew up in Hnifsdalur, a fishing village of three hundred people in northwest Iceland, where his father was a carpenter, his forebears fishermen. He is a large-boned lumberjack of a man with deep-blue eyes and dark, swept-back hair. During the Vestmannaeyjar eruption, his varied professorial commitments caused him now and again to disappear. After the first three weeks of attacking the lava with the new invasion pumps—during the early part of which he slept three hours a night—he went off to Penn State, where he led a seminar called Volcano in Town.

As each new road was built on the lava, it would move and be replaced, move and be replaced, until the lateral motion stopped. In Thorbjorn's words, "You could see that the lava was stopping, and you could extend the pipes. We moved closer to the crater, moved gradually up there—from place to place, a week at a time, just letting water pour out on the lava. If you look near the volcano, the surface there now is extremely rough. Close to the crater, the flow movement was fast. With the cooling, you create huge blocks of solid matter that move on the lava stream and turn over." When ice forms in the Yukon River, it makes huge blocks that move on the stream and turn over. Eventually, they jam—piled one upon another in a stilled surface that is mountainous and rough.

If the Easter Flow had gone more north than east, it would have overwhelmed all opposition, ending the battle in the absolute defeat of people and pumps. Instead, it filled a huge space but damaged nothing. Possibly, or probably, it was diverted.

"It was enough lava to cover the town and fill the harbor," Thorbjorn said. "I tend to think that it went that way because the lava in front of the crater was more stable."

The lava in front of the crater was more stable because it was in the process of being bathed in six million tons of water. As Gudmundur Karlsson said, "I am convinced that because we were pumping and this part of the lava was cooled most of the new flow went east and out to sea. If we hadn't done something, I very much doubt that we would be here now. Of course, we can't do anything if the natural force is strong, but I am convinced that this really helped. We tried, anyway."

They cooled (and hardened) more than five million cubic yards of lava. They left upon it two hundred thousand tons of salt.

On July 3, five and a half months after the fissure opened

and the fire curtains rose, Thorbjorn went down into the crater of the new volcano and pronounced the eruption dead—a remarkably decisive thing for a scientist to do. Thorbjorn felt it was important. He explains, "The production had decreased very regularly. It was almost possible to extrapolate and see when it would stop. The day we went down into the crater, it was completely stationary and cool. The air was cool there. It was very peaceful. We issued a proclamation that the eruption had stopped. We took the risk of telling people that the eruption was over, so they could come out and settle down again and build up the town." In the National Emergency Operation Center, the Vestmannaeyjar campaign is remembered in the glow of total triumph. As Thorbjorn is the first to suggest, however, the true extent of the victory will never be known—the role of luck being unassessable, the effects of intervention being ultimately incalculable, and the assertion that people can stop a volcano being hubris enough to provoke a new eruption.

Los Angeles

Against the Mountains

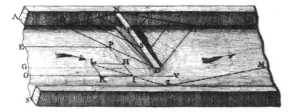

IN LOS ANGELES versus the San Gabriel Mountains, it is not always clear which side is losing. For example, the Genofiles, Bob and Jackie, can claim to have lost and won. They live on an acre of ground so high that they look across their pool and past the trunks of big pines at an aerial view over Glendale and across Los Angeles to the Pacific bays. The setting, in cool dry air, is serene and Mediterranean. It has not been everlastingly serene.

On a February night some years ago, the Genofiles were awakened by a crash of thunder—lightning striking the mountain front. Ordinarily, in their quiet neighborhood, only the creek beside them was likely to make much sound, dropping steeply out of Shields Canyon on its way to the Los Angeles River. The creek, like every component of all the river systems across the city from mountains to ocean, had not been left to nature. Its banks were concrete. Its bed was concrete. When boulders were running there, they sounded like a rolling freight. On a night like this, the boulders should have been running. The creek should have been a torrent. Its unnatural sound was unnaturally absent. There was, and had been, a lot of rain.

The Genofiles had two teen-age children, whose rooms were on the uphill side of the one-story house. The window in Scott's room looked straight up Pine Cone Road, a cul-de-sac, which, with hundreds like it, defined the northern limit of the city, the confrontation of the urban and the wild. Los Angeles is overmatched on one side by the Pacific Ocean and on the other by very high mountains. With respect to these principal boundaries, Los Angeles is done sprawling. The San Gabriels, in their state of tectonic youth, are rising as rapidly as any range on earth. Their loose inimical slopes flout the tolerance of the angle of repose. Rising straight up out of the megalopolis, they stand ten thousand feet above the nearby sea, and they are not kidding with this city. Shedding, spalling, self-destructing, they are disintegrating at a rate that is also among the fastest in the world. The phalanxed communities of Los Angeles have pushed themselves hard against these mountains, an aggression that requires a deep defense budget to contend with the results. Kimberlee Genofile called to her mother, who joined her in Scott's room as they looked up the street. From its high turnaround, Pine Cone Road plunges downhill like a ski run, bending left and then right and then left and then right in steep christiania turns for half a mile above a three-hundred-foot straightaway that aims directly at the Genofiles' house. Not far below the turnaround, Shields Creek passes under the street, and there a kink in its concrete profile had been plugged by a six-foot boulder. Hence the silence of the creek. The water was now spreading over the street. It descended in heavy sheets. As the young Genofiles and their mother glimpsed it in the all but total darkness, the scene was suddenly illuminated by a blue electrical flash. In the blue light they saw a massive blackness, moving. It was not a landslide, not a mudslide, not a rock avalanche; nor by any means was it the front of a conventional flood. In Jackie's words, "It was just one big black thing coming at us, rolling, rolling with a lot of water in front of it, pushing the water, this big black thing. It was just one big black hill coming toward us."

In geology, it would be known as a debris flow. Debris flows amass in stream valleys and more or less resemble fresh concrete. They consist of water mixed with a good deal of solid material, most of which is above sand size. Some of it is Chevrolet size. Boulders bigger than cars ride long distances in debris flows. Boulders grouped like fish eggs pour downhill in debris flows. The dark material coming toward the Genofiles was not only full of boulders; it was so full of automobiles it was like bread dough mixed with raisins. On its way down Pine Cone Road, it plucked up cars from driveways and the street. When it crashed into the Genofiles' house, the shattering of safety glass made terrific explosive sounds. A door burst open. Mud and boulders poured into the hall. We're going to go, Jackie thought. Oh, my God, what a hell of a way for the four of us to die together.

The parents' bedroom was on the far side of the house. Bob Genofile was in there kicking through white satin draperies at the panelled glass, smashing it to provide an outlet for water, when the three others ran in to join him. The walls of the house neither moved nor shook. As a general contractor, Bob had built dams, department stores, hospitals, six schools, seven churches, and this house. It was made of concrete block with steel reinforcement, sixteen inches on center. His wife had said it was stronger than any dam in California. His crew had called it "the fort." In those days, twenty years before, the Genofiles' acre was close by the edge of the mountain brush, but a developer had come along since then and knocked down thousands of trees and put Pine Cone Road up the slope. Now Bob Genofile was thinking, I hope the roof holds. I hope the roof is strong enough to hold. Debris was flowing over it. He told Scott to shut the bedroom door. No sooner was the door closed than it was battered down and fell into the room. Mud, rock, water poured in. It pushed everybody against the far wall. "Jump on the bed," Bob said. The bed began to rise. Kneeling on it—on a gold velvet spread—they could soon press their palms against the ceiling. The bed also moved toward the glass wall. The two teen-agers got off, to try to control the motion, and were pinned between the bed's brass railing and the wall. Boulders went up against the railing, pressed it into their legs, and held them fast. Bob dived into the muck to try to move the boulders, but he failed. The debris flow, entering through windows as well as doors, continued to rise. Escape was still possible for the parents but not for the children. The parents looked at each other and did not stir. Each reached for and held one of the children. Their mother felt suddenly resigned, sure that her son and daughter would die and she and her husband would quickly follow. The house became buried to the eaves. Boulders sat on the roof. Thirteen automobiles were packed around the building, including five in the pool. A din of rocks kept banging against them. The stuck horn of a buried car was blaring. The family in the darkness in their fixed tableau watched one another by the light of a directional signal, endlessly blinking. The house had filled up in six minutes, and the mud stopped rising near the children's chins.

STORIES LIKE THAT do not always have such happy endings. A man went outside to pick up his newspaper one morning, heard a sound, turned, and died of a heart attack as he saw his house crushed to pieces with his wife and two children inside. People have been buried alive in their beds. But such cases arc infrequent. Debris flows generally are much less destructive of life than of property. People get out of the way.

If they try to escape by automobile, they have made an obvious but imperfect choice. Norman Reid backed his Pontiac into the street one January morning and was caught from behind by rock porridge. It embedded the car to the chrome strips. Fifty years of archival news photographs show cars of every vintage standing like hippos in chunky muck. The upper halves of their headlights peep above the surface. The late Roland Case Ross, an emeritus professor at California State University, told me of a day in the early thirties when he watched a couple rushing to escape by car. She got in first. While her husband was going around to get in his side, she got out and ran into the house for more silverware. When the car at last putt-putted downhill, a wall of debris was nudging the bumper. The debris stayed on the vehicle's heels all the way to Foothill Boulevard, where the car turned left.

Foothill Boulevard was U.S. Route 66—the western end of the rainbow. Through Glendora, Azusa, Pasadena, it paralleled the mountain front. It strung the metropolitan border towns. And it brought in emigrants to fill them up. The real-estate line of maximum advance now averages more than a mile above Foothill, but Foothill receives its share of rocks. A debris flow that passed through the Monrovia Nursery went on to Foothill and beyond. With its twenty million plants in twelve hundred varieties, Monrovia was the foremost container nursery in the world, and in its recovery has remained so. The debris flow went through the place picking up pots and cans. It got into a greenhouse two hundred feet long and smashed out the southern wall, taking bougainvillea and hibiscus with it. Arby's, below Foothill, blamed the nursery for damages, citing the hibiscus that had come with the rocks. Arby's sought compensation, but no one was buying beef that thin.

In the same storm, large tree trunks rode in the debris like javelins and broke through the sides of houses. Automobiles went in through picture windows. A debris flow hit the gym at Azusa Pacific College and knocked a large hole in the upslope wall. In the words of Cliff Hamlow, the basketball coach, "If we'd had students in there, it would have killed them. Someone said it sounded like the roar of a jet engine. It filled the gym up with mud, and with boulders two and three feet in diameter. It went out through the south doors and spread all over the football field and track. Chain-link fencing was sheared off—like it had been cut with a welder. The place looked like a war zone." Azusa Pacific College wins national championships in track, but Coach Hamlow's basketball team (12–18) can't get the boulders out of its game.

When a debris flow went through the Verdugo Hills Cemetery, which is up a couple of switchbacks on the mountain front, two of the central figures there, resting under impressive stones, were "Hiram F. Hatch, 1st Lieut. 6th Mich. Inf., December 24, 1843–October 12, 1922," and "Henry J. Hatch, Brigadier General, United States Army, April 28, 1869–December 31, 1931." The two Hatches held the hill while many of their comrades slid below. In all, thirty-five coffins came out of the cemetery and took off for lower ground. They went down Hillrose Street and were scattered over half a mile. One came to rest in the parking lot of a supermarket. Many were reburied by debris and, in various people's yards, were not immediately found. Three turned up in one yard. Don Sulots, who had moved into the fallout path two months before, said, "It sounded like thunder. By the time I made it to the front door and got it open, the muck was already three feet high. It's quite a way to start off life in a new home—mud, rocks, and bodies all around."

Most people along the mountain front are about as mindful of debris flows as those corpses were. Here today, gone tomorrow. Those who worry build barricades. They build things called deflection walls—a practice that raises legal antennae and, when the caroming debris breaks into the home of a neighbor, probes the wisdom of Robert Frost. At least one family has experienced so many debris flows coming through their back yard that they long ago installed overhead doors in the rear end of their built-in garage. To guide the flows, they put deflection walls in their back yard. Now when the boulders come they open both ends of their garage, and the debris goes through to the street.

Between Harrow Canyon and Englewild Canyon, a private street called Glencoe Heights teased the mountain front. Came a time of unprecedented rain, and the neighborhood grew ever more fearful—became in fact so infused with catastrophic anticipation that it sought the drastic sort of action that only a bulldozer could provide. A fire had swept the mountainsides, leaving them vulnerable, dark, and bare. Expecting floods of mud and rock, people had piled sandbags and built heavy wooden walls. Their anxiety was continuous for many months. "This threat is on your mind all the time," Gary Lukehart said. "Every time you leave the house, you stop and put up another sandbag, and you just hope everything will be all right when you get back." Lukehart was accustomed to losing in Los Angeles. In the 1957 Rose Bowl, he was Oregon State's quarterback. A private street could not call upon city or county for the use of heavy equipment, so in the dead of night, as steady rain was falling, a call was put in to John McCafferty—bulldozer for hire. McCafferty had a closeup knowledge of the dynamics of debris flows: he had worked the mountain front from San Dimas to Sierra Madre, which to him is Sarah Modri. ("In those canyons at night, you could hear them big boulders comin'. They sounded like thunder.") He arrived at Glencoe Heights within the hour and set about turning the middle of the street into the Grand Canal of Venice. His Cat was actually not a simple dozer but a 955 loader on tracks, with a two-and-a-quarter-yard bucket seven feet wide. Cutting water mains, gas mains, and sewers, he made a ditch that eventually extended five hundred feet and was deep enough to take in three thousand tons of debris. After working for five hours, he happened to be by John Caufield's place ("It had quit rainin', it looked like the worst was over") when Caufield came out and said, "Mac, you sure have saved my bacon."

McCafferty continues, "All of a sudden, we looked up at the mountains—it's not too far from his house to the mountains, maybe a hundred and fifty feet—and we could just see it all comin'. It seemed the whole mountain had come loose. It flowed like cement." In the ditch, he put the Cat in reverse and backed away from the oncoming debris. He backed three hundred feet. He went up one side of the ditch and was about halfway out of it when the mud and boulders caught the Cat and covered it over the hood. In the cab, the mud pushed against McCafferty's legs. At the same time, debris broke into

Caufield's house through the front door and the dining-room window, and in five minutes filled it to the eaves.

Other houses were destroyed as well. A garage left the neighborhood with a car in it. One house was buried twice. (After McCafferty dug it out, it was covered again.) His ditch, however, was effective, and saved many places on slightly higher ground, among them Gary Lukehart's and the home of John Marcellino, the chief executive officer of Mackinac Island Fudge. McCafferty was promised a lifetime supply of fudge. He was on the scene for several days, and in one span worked twenty-four hours without a break. The people of the street brought him chocolate milkshakes. He had left his lowbed parked around the corner. When at last he returned to it and prepared to go home, he discovered that a cop had given him a ticket.

A METROPOLIS that exists in a semidesert, imports water three hundred miles, has inveterate flash floods, is at the grinding edges of two tectonic plates, and has a microclimate tenacious of noxious oxides will have its priorities among the aspects of its environment that it attempts to control. For example, Los Angeles makes money catching water. In a few days in 1983, it caught twenty-eight million dollars' worth of water. In one period of twenty-four hours, however, the ocean hit the city with twenty-foot waves, a tornado made its own freeway, debris flows poured from the San Gabriel front, and an earthquake shook the region. Nature's invoice was forty million dollars. Later, twenty million more was spent dealing with the mountain debris.

There were those who would be quick—and correct—in

saying that were it not for the alert unflinching manner and imaginative strategies by which Los Angeles outwits the mountains, nature's invoices at such times would run into the billions. The rear-guard defenses are spread throughout the city and include more than two thousand miles of underground conduits and concrete-lined open stream channels—a web of engineering that does not so much reinforce as replace the natural river systems. The front line of battle is where the people meet the mountains—up the steep slopes where the subdivisions stop and the brush begins.

Strung out along the San Gabriel front are at least a hundred and twenty bowl-shaped excavations that resemble football stadiums and are often as large. Years ago, when a big storm left back yards and boulevards five feet deep in scree, one neighborhood came through amazingly unscathed, because it happened to surround a gravel pit that had filled up instead. A tungsten filament went on somewhere above Los Angeles. The county began digging pits to catch debris. They were quarries, in a sense, but exceedingly bizarre quarries, in that the rock was meant to come to them. They are known as debris basins. Blocked at their downstream ends with earthfill or concrete constructions, they are also known as debris dams. With clean spillways and empty reservoirs, they stand ready to capture rivers of boulders—these deep dry craters, lying close above the properties they protect. In the overflowing abundance of urban nomenclature, the individual names of such basins are obscure, until a day when they appear in a headline in the Los Angeles *Times*: Harrow, Englewild, Zachau, Dunsmuir, Shields, Big Dalton, Hog, Hook East, Hook West, Limekiln, Starfall, Sawpit, Santa Anita. For fifty miles, they mark the wild boundary like bulbs beside a mirror. Behind chain links, their idle ovate forms more than suggest defense.

They are separated, on the average, by seven hundred yards. In aggregate, they are worth hundreds of millions of dollars. All this to keep the mountains from falling on Johnny Carson.

The principal agency that developed the debris basins was the hopefully named Los Angeles County Flood Control District, known familiarly through the region as Flood Control, and even more intimately as Flood. ("When I was at Flood, one of our dams filled with debris overnight," a former employee remarked to me. "If any more rain came, we were going to have to evacuate the whole of Pasadena.") There has been a semantic readjustment, obviously intended to acknowledge that when a flood pours out of the mountains it might be half rock. The debris basins are now in the charge of the newly titled Sedimentation Section of the Hydraulic Division of the Los Angeles County Department of Public Works. People still call it Flood. By whatever name the agency is called, its essential tactic remains unaltered. This was summarized for me in a few words by an engineer named Donald Nichols, who pointed out that eight million people live below the mountains on the urban coastal plain, within an area large enough to accommodate Philadelphia, Detroit, Chicago, St. Louis, Boston, and New York. He said, "To make the area inhabitable, you had to put in lined channels on the plain and halt the debris at the front. If you don't take it out at the front, it will come out in the plain, filling up channels. A filled channel won't carry diddly-boo."

To stabilize mountain streambeds and stop descending rocks even before they reach the debris basins, numerous crib structures (barriers made of concrete slats) have been emplaced in high canyons—the idea being to convert plunging streams into boulder staircases, and hypothetically cause erosion to work against itself. Farther into the mountains, a dozen dams of some magnitude were built in the nineteen-twenties and thirties to control floods and conserve water. Because they are in the San Gabriels, they inadvertently trap large volumes of debris. One of them—the San Gabriel Dam, in the San Gabriel River—was actually built as a debris-control structure. Its reservoir, which is regularly cleaned out, contained, just then, twenty million tons of mountain.

The San Gabriel River, the Los Angeles River, and the Big Tujunga (Bigta Hung-ga) are the principal streams that enter the urban plain, where a channel that filled with rock wouldn't carry diddly-boo. Three colossal debris basins—as diffcrent in style as in magnitude from those on the mountain front—have been constructed on the plain to greet these rivers. Where the San Gabriel goes past Azusa on its way to Alamitos Bay, the Army Corps of Engineers completed in the late nineteen-forties a dam ninety-two feet high and twenty-four thousand feet wide—this to stop a river that is often dry, and trickles most of the year. Santa Fe Dam, as it is called, gives up at a glance its own story, for it is made of boulders that are shaped like potatoes and are generally the size of watermelons. They imply a large volume of water flowing with high energy. They are stream-propelled, stream-rounded boulders, and the San Gabriel is the stream. In Santa Fe Basin, behind the dam, the dry bed of the San Gabriel is half a mile wide. The boulder-strewn basin in its entirety is four times as wide as that. It occupies eighteen hundred acres in all, nearly three square miles, of what would be prime real estate were it not for the recurrent arrival of rocks. The scene could have been radioed home from Mars, whose cobbly face is in part the result of debris flows dating to a time when Mars had surface water.

The equally vast Sepulveda Basin is where Los Angeles receives and restrains the Los Angeles River. In Sepulveda

Basin are three golf courses, which lend ample support to the widespread notion that everything in Los Angeles is disposable. Advancing this national prejudice even further, debris flows, mudslides, and related phenomena have "provided literary minds with a ready-made metaphor of the alleged moral decay of Los Angeles." The words belong to Reyner Banham, late professor of the history of architecture at University College, London, whose passionate love of Los Angeles left him without visible peers. The decay was only "alleged," he said. Of such nonsense he was having none. With his "Los Angeles: The Architecture of Four Ecologies," Banham had become to this deprecated, defamed, traduced, and disparaged metropolis what Pericles was to Athens. Banham knew why the basins were there and what the people were defending. While all those neurasthenic literary minds are cowering somewhere in ethical crawl space, the quality of Los Angeles life rises up the mountain front. There is air there. Cool is the evening under the crumbling peaks. Cool descending air. Clean air. Air with a view. "The financial and topographical contours correspond almost exactly," Banham said. Among those "narrow, tortuous residential roads serving precipitous house-plots that often back up directly on unimproved wilderness" is "the fat life of the delectable mountains."

People of Gardena, Inglewood, and Watts no less than Azusa and Altadena pay for the defense of the mountain front, the rationale being that debris trapped near its source will not move down and choke the channels of the inner city, causing urban floods. The political City of Los Angeles—in its vague and tentacular configuration—actually abuts the San Gabriels for twenty miles or so, in much the way that it extends to touch the ocean in widely separated places like Venice, San Pedro, and Pacific Palisades. Los Angeles County reaches across the mountains and far into the Mojave Desert. The words "Los Angeles" as generally used here refer neither to the political city nor to the county but to the multinamed urban integrity that has a street in it seventy miles long (Sepulveda Boulevard) and, from the Pacific Ocean at least to Pomona, moves north against the mountains as a comprehensive town.

The debris basins vary greatly in size—not, of course, in relation to the populations they defend but in relation to the watersheds and washes above them in the mountains. For the most part, they are associated with small catchments, and the excavated basins are commensurately modest, with capacities under a hundred thousand cubic yards. In a typical empty reservoir—whatever its over-all dimensions may be—stands a columnar tower that resembles a campanile. Full of holes, it is known as a perforated riser. As the basin fills with a thick-flowing slurry of water, mud, and rock, the water goes into the tower and is drawn off below. The county calls this water harvesting.

Like the freeways, the debris-control system ordinarily functions but occasionally jams. When the Genofiles' swimming pool filled with cars, debris flows descended into other neighborhoods along that part of the front. One hit a culvert, plugged the culvert, crossed a road in a bouldery wave, flattened fences, filled a debris basin, went over the spillway, and spread among houses lying below, shoving them off their foundations. The debris basins have caught as much as six hundred thousand cubic yards in one storm. Over time, they have trapped some twenty million tons of mud and rock. Inevitably, sometimes something gets away.

At Devils Gate—just above the Rose Bowl, in Pasadena— a dam was built in 1920 with control of water its only objective.

Yet its reservoir, with a surface of more than a hundred acres, has filled to the brim with four million tons of rock, gravel, and sand. A private operator has set up a sand-and-gravel quarry in the reservoir. Almost exactly, he takes out what the mountains put in. As one engineer has described it, "he pays Flood, and Flood makes out like a champ."

IT WAS ASSUMED that the Genofiles were dead. Firemen and paramedics who came into the neighborhood took one glance at the engulfed house and went elsewhere in search of people needing help. As the family remained trapped, perhaps an hour went by. They have no idea.

"We didn't know why it had come or how long it was going to last."

They lost all sense of time. The stuck horn went on blaring, the directional signal eerily blinking. They imagined that more debris was on the way.

"We didn't know if the whole mountain was coming down."

As they waited in the all but total darkness, Jackie thought of neighbors' children. "I thought, Oh, my gosh, all those little kids are dead. Actually, they were O.K. And the neighbors thought for sure we were all gone. All our neighbors thought we were gone."

At length, a neighbor approached their house and called out, "Are you alive?"

"Yes. But we need help."

As the debris flow hit the Genofiles' house, it also hit a six-ton truck from the L.A.C.F.C.D., the vigilant bureau called Flood. Vigilance was about all that the L.A.C.F.C.D. had been able to offer. The patrolling vehicle and its crew of two were as helpless as everyone else. Each of the crewmen had lived twenty-six years, and each came close to ending it there. Minutes before the flow arrived, the truck labored up Pine Cone Road—a forty-one-per-cent grade, steep enough to stiff a Maserati. The two men meant to check on a debris basin at the top. Known as Upper Shields, it was less than two years old, and had been built in anticipation of the event that was about to occur. Oddly enough, the Genofiles and their neighbors were bracketed with debris basins—Upper Shields above them, Shields itself below them, six times as large. Shields Debris Basin, with its arterial concrete feeder channels, was prepared to catch fifty thousand tons. The Genofiles' house looked out over Shields as if it were an empty lake, its shores hedged about with oleander. When the developer extended Pine Cone Road up into the brush, the need for Upper Shields was apparent. The new basin came in the nick of time but—with a capacity under six thousand cubic yards—not in the nick of space. Just below it was a chain-link gate. As the six-ton truck approached the gate, mud was oozing through. The basin above had filled in minutes, and now, suddenly, boulders shot like cannonballs over the crest of the dam, with mud, cobbles, water, and trees. Chris Terracciano, the driver, radioed to headquarters, "It's coming over." Then he whipped the truck around and fled. The debris flow came through the chain-link barrier as if the links were made of paper. Steel posts broke off. As the truck accelerated down the steep hill, the debris flow chased and caught it. Boulders bounced against it. It was hit by empty automobiles spinning and revolving in the muck. The whole descending complex gathered force with distance. Terracciano later said, "I thought I was dead the whole way." The truck finally stopped when it bashed against a tree and a cement-block wall. The rear window shattered.

Terracciano's partner suffered a broken leg. The two men crawled out through the window and escaped over the wall.

Within a few miles, other trapped patrols were calling in to say, "It's coming over." Zachau went over—into Sunland. Haines went over—into Tujunga. Dunsmuir went over—into Highway Highlands. As bulldozers plow out the streets after events like these, the neighborhoods of northern Los Angeles assume a macabre resemblance to New England villages under deep snow: the cleared paths, the vehicular rights-of-way, the parking meters buried within the high banks, the half-covered drift-girt homes. A street that is lined with palms will have debris berms ten feet up the palms. In the Genofiles' front yard, the drift was twelve feet deep. A person, without climbing, could walk onto the roof. Scott's bedroom had a few inches of space left at the top. Kimberlee's had mud on the ceiling. On the terrace, the crushed vehicles, the detached erratic wheels suggested bomb damage, artillery hits, the track of the Fifth Army. The place looked like a destroyed pillbox. No wonder people assumed that no one had survived inside.

There was a white sedan under the house eaves crushed to half its height, with two large boulders resting on top of it. Near the pool, a Volkswagen bug lay squashed. Another car was literally wrapped around a tree, like a C-clamp, its front and rear bumpers pointing in the same direction. A crushed pickup had boulders all over it, each a good deal heavier than anything a pickup could carry. One of the cars in the swimming pool was upside down, its tires in the air. A Volkswagen was on top of it. Bob Genofile—owner, contractor, victim— walked around in rubber boots, a visored construction cap, a foul-weather jacket, studying the damage, mostly guessing at what he couldn't see. A big, strongly built, leonine man with prematurely white hair, he looked like a middle linebacker near the end of a heavy day. He wondered if the house was still on its foundation, but there was no telling in this profound chaos, now hardening and cracking like bad concrete. In time, as his house was excavated from the inside, he would find that it had not budged. Not one wall had so much as cracked. He was uninsured, but down in the rubble was a compensation of greater value than insurance. Forever, he could say, as he quietly does when he tells the story, "I built it, man."

Kimberlee's birthday came two days after the debris. She was a college student, turning nineteen, and her father had had a gift for her that he was keeping in his wallet. "I had ninetcen fifty-dollar bills to give her for her birthday, but my pants and everything was gone."

Young Scott, walking around in the wreckage, saw a belt sticking out of the muck like a night crawler after rain. He pulled at it, and the buried pants came with it. The wallet was still in the pants. The wallet still contained what every daughter wants for her birthday: an album of portraits of U.S. Grant, no matter if Ulysses is wet or dry.

The living room had just been decorated, and in six minutes the job had been destroyed—"the pale tangerines and greens, Italian-style furniture with marble, and all that." Jackie Genofile continues the story: "We had been out that night, and, you know, you wear your better jewelry. I came home like an idiot and put mine on the dresser. Bob put his on the dresser. Three weeks later, when some workers were cleaning debris out of the bedroom, they found his rings on the floor. They did not find mine. But—can you believe it?—a year and a half later Scott was down in the debris basin with one of his friends, and the Flood Control had these trucks there cleaning it out, and Scott saw this shiny thing, and he picked it up, and it was my ring that Bob had given me just before the storm."

Before the storm, they had not in any way felt threatened. Like their neighbors, they were confident of the debris basins, of the concrete liners of the nearby stream. After the storm, neighbors moved away. Where Pine Cone Road swung left or right, the debris had made centrifugal leaps, breaking into houses. A hydrant snapped off, and arcing water shot through an upstairs window. A child nearly drowned inside his own house. The family moved. "Another family that moved owned one of the cars that ended up in our pool," Jackie told me. "The husband said he'd never want to live here again, you know. And she was in real estate."

After the storm, the Genofiles tended to wake in the night, startled and anxious. They still do. "I wake up once in a while really uptight," Bob said. "I can just feel it—go through the whole thing, you know."

Jackie said that when rain pounds on a roof, anywhere she happens to be, she will become tense. Once, she took her dog and her pillow and went to sleep in Bob's office—which was then in Montrose, down beyond Foothill Boulevard.

Soon after the storm, she said, "Scotty woke up one night, and he had a real high temperature. You see, he was sixteen, and he kept hearing the mud and rock hitting the window. He kept thinking it was going to come again. Kim used to go four-wheeling, and cross streams, and she had to get out once, because they got stuck, and when she felt the flow of water and sand on her legs, she said, she could have panicked."

Soon after the storm, the family gathered to make a decision. Were they going to move or were they going to dig out their house and rebuild it? Each of them knew what might have happened. Bob said, "If it had been a frame house, we would be dead down in the basin below."

But it was not a frame house. It was the fort. "The kids said rebuild. So we rebuilt."

As he sat in his new living room telling the story, Bob was dressed in a Pierre Cardin jumper and pants, and Jackie was beside him in a pale-pink jumpsuit by Saint Germain. The house had a designer look as well, with its railings and balconies and Italianate marbles under the tall dry trees. It appeared to be worth a good deal more than the half-million dollars Bob said it might bring. He had added a second story and put all bedrooms there. The original roof spreads around them like a flaring skirt. He changed a floor-length window in the front hall, filling the lower half of it with cement block.

I asked what other structural changes he had made.

He said, "None."

The Genofiles sued Los Angeles County. They claimed that Upper Shields Debris Basin had not been cleaned out and that the channel below was improperly designed. Los Angeles settled for three hundred and thirty-seven thousand five hundred dollars.

From the local chamber of commerce the family later received the Beautification Award for Best Home. Two of the criteria by which houses are selected for this honor are "good maintenance" and "a sense of drama."

I HAVE NOT BEEN specific about the dates of the stories so far recounted. This was to create the impression that debris pours forth from the mountains continually, perennially, perpetually—which it does and does not, there being a great temporal disparity between the pace at which the mountains behave and the way people think. Debris flows do not occur in every possible season. When they do happen, they don't just spew from any canyon but come in certain places on the mountain front. The places change. Volumes differ. There are vintage years. The four most prominent in this century have been 1934, 1938, 1969, and 1978. Exceptional flows have occurred at least once a decade, and lesser ones in greater numbers. Exceptional flows are frequent, in other words, but not frequent enough to deter people from building pantiled mansions in the war zone, dingbats in the line of fire.

Why the debris moves when it does or where it does is not attributable to a single agent. The parent rock has been extensively broken up by earthquakes, but that alone will not make it flow. Heavy rainfall, the obvious factor, is not as obvious as it may seem. In 1980, some of the most intense storms ever measured in Los Angeles failed to produce debris flows of more than minimal size. The setting up of a debris flow is a little like the charging of an eighteenth-century muzzle-loader: the ramrod, the powder, the wadding, the shot. Nothing much would happen in the absence of any one component. In sequence and proportion each had to be correct.

On the geologic time scale, debris flows in the San Gabriel Mountains can be looked upon as constant. With all due respect, though, the geologic time scale doesn't mean a whole lot in a place like Los Angeles. In Los Angeles, even the Los Angeles time scale does not arouse general interest. A super-event in 1934? In 1938? In 1969? In 1978? Who is going to remember that? A relatively major outpouring—somewhere in fifty miles—about once every decade? Mountain time and city time appear to be bifocal. Even with a geology functioning at such remarkably short intervals, the people have ample time to forget it.

In February of 1978, while debris was still hardening in the home of the Genofiles, Wade Wells, of the United States Forest Service, went up and down Pine Cone Road knocking on doors, asking how long the people had lived there. He wondered who remembered, nine years back, the debris-flow inundations of Glendora and Azusa, scarcely twenty miles away. Only two did. Everyone else had arrived since 1969.

Wells is a hydrologist who works in the mountains, principally in San Dimas Experimental Forest, where he does research on erosion and sedimentation—the story of assembling debris. With a specialist's eye, he notes the mountain front, and in its passivity can see the tension: "These guys here, they should be nervous when it rains. Their houses are living on borrowed time. See that dry ledge? It's a waterfall. I've seen hundreds of tons of rock falling over it." More often, though, he is thousands of feet above the nearest house, on slopes so steep he sometimes tumbles and rolls. With his colleagues, he performs experiments with plants, rock, water, fire. When I first became interested in Los Angeles' battle with debris flows, I went up there with them a number of times. The mountains, after all, are where the rocks come from. The mountains shape the charge that will advance upon the city. People come from odder places than the East Coast to see this situation. One day, a couple of scientists arrived from the Cordillera Cantábrica, in northwestern Spain. When they saw how rapidly the San Gabriels were disintegrating, one of them said he felt sorry for Wells, who would soon be out of work. When Wells told him that the mountains were rising even faster than they were coming down, the man said, *"Muy interesante. Sí, señor."*

From below, one look at the San Gabriels will suggest their advantage. The look is sometimes hard to come by. You might be driving up the San Gabriel River Freeway in the morning, heading straight at the mountains at point-blank range, and not be able to see them. A voice on KNX tells you that the day is clear. There's not a cloud in the sky, as the blue straight up confirms. A long incline rises into mist, not all of which is smog. From time immemorial, this pocket of the coast has been full of sea fog and persistent vapors. The early Spaniards called it the Bay of Smokes. Smog, the action of sunlight on nitrogen oxides, has only contributed to a preexisting veil. Sometimes you don't see the San Gabriels until the streets stop and the mountains start. The veil suddenly thins, and there they are, in height and magnitude overwhelming. You plunge into a canyon flanked with soaring slopes before you realize you are out of town. The San Gabriel Mountains are as rugged as any terrain in America, and their extraordinary proximity to the city, the abruptness of the transition from the one milieu to the other, cannot be exaggerated. A lone hiker in the San Gabriels one winter—exhausted, snow-blinded, hypothermic—staggered down a ridgeline out of the snow and directly into the parking lot of a shopping center, where he crawled to a phone booth, called 911, and slumped against the glass until an ambulance came to save him.

Hang-glider pilots go up the San Gabriels, step off crags, and, after a period of time proportional to their skills, land somewhere in the city. The San Gabriels are nearly twice as high as Mt. Katahdin or Mt. Washington, and are much closer to the sea. From base platform to summit, the San Gabriels are three thousand feet higher than the Rockies. To be up in the San Gabriels is to be both above and beside urban Los Angeles, only minutes from the streets, and to look north from ridge to dry ridge above deeply cut valleys filled with gulfs of clear air. Beyond the interior valleys—some fifty thousand feet

away and a vertical mile above you—are the summits of Mt.
Baldy, Mt. Hawkins, Mt. Baden-Powell. They are so clearly
visible in the dry blue sky that just below their ridgelines you
can almost count the boulders that are bunched there like
grapes.

If you turn and face south, you look out over something
like soft slate that reaches fifty miles to an imprecise horizon.
The whole of Los Angeles is spread below you, and none of
it is visible. It is lost absolutely in the slate-gray sea, grayer
than a minesweeper, this climatic wonder, this megalopolitan
featherbed a thousand feet thick, known as "the marine layer."
Early in the day, it is for the most part the natural sea fog. As
you watch it from above through the morning and into the
afternoon, it turns yellow, and then ochre, and then brown,
and sometimes nearly black—like butter darkening in a skillet.

Glancing down at it one day while working on an experiment, Wade Wells said it seemed to have reached the hue
of a first-stage smog alert. Wells was helping Edwin Harp, a
debris-flow specialist from the United States Geological Survey, collect "undisturbed" samples by hammering plastic tubes
into the mountain soil.

"If the soil were nice and compliant, this would be nice
and scientific," Harp said, smacking the plastic with a wooden-
handled shovel. After a while, he extracted a tube full of
uncompliant material, and said, "This isn't soil; it's regolith."
Regolith is a stony blanket that lies under soil and over bedrock.
It crumbled and was pebbly in the hand.

As they prepared to sink another tube, I said, "What's a
first-stage smog alert?"

"Avoid driving, avoid strenuous activity," Wells answered.

Harp said, "Avoid breathing."

The slope they were sampling had an incline of eighty-five per cent. They were standing, and walking around, but I preferred—just there—to sit. Needle grass went through my trousers. The heads of needle grass detach from the stalks and have the barbed design of arrows. They were going by the quiver into my butt but I still preferred to sit. It was the better posture for writing notes. The San Gabriels are so steep and so extensively dissected by streams that some watersheds are smaller than a hundred acres. The slopes average sixty-five to seventy per cent. In numerous places, they are vertical. The angle of repose—the steepest angle that loose rocks can abide before they start to move, the steepest angle the soil can maintain before it starts to fail—will vary locally according to the mechanics of shape and strength. Many San Gabriel slopes are at the angle of repose or beyond it. The term "oversteepened" is often used to describe them. At the giddy extreme of oversteepening is the angle of maximum slope. Very large sections of the San Gabriels closely approach that angle. In such terrain, there is not much to hold the loose material except the plants that grow there.

Evergreen oaks were fingering up the creases in the mountainsides, pointing toward the ridgeline forests of big-cone Douglas fir, of knobcone and Coulter pine. The forests had an odd sort of timberline. They went down to it rather than up. Down from the ridges the conifers descended through nine thousand, seven thousand, six thousand feet, stopping roughly at five. The forests abruptly ended—the country below being too dry in summer to sustain tall trees. On down the slopes and all the way to the canyons was a thicket of varied shrubs that changed in character as altitude fell but was everywhere dense enough to stop an army. On its lower levels, it was all green, white, and yellow with buckwheat, burroweed, lotus and sage, deerweed, bindweed, yerba santa. There were wild morning glories, Canterbury bells, tree tobacco, miner's lettuce. The thicket's resistance to trespass, while everywhere formidable, stiffened considerably as it evolved upward. There were intertwining mixtures of manzanita, California lilac, scrub oak, chamise. There was buckthorn. There was mountain mahogany. Generally evergreen, the dark slopes were splashed here and there with dodder, its mustard color deepening to rust. Blossoms of the Spanish bayonet stood up like yellow flames. There were lemonade berries (relatives of poison ivy and poison oak). In canyons, there were alders, big-leaf-maple bushes, pug sycamores, and California bay. Whatever and wherever they were, these plants were prickly, thick, and dry, and a good deal tougher than tundra. Those evergreen oaks fingering up the creases in the mountains were known to the Spaniards as chaparros. Riders who worked in the related landscape wore leather overalls open at the back, and called them chaparajos. By extension, this all but impenetrable brush was known as chaparral.

The low stuff, at the buckwheat level, is often called soft chaparral. Up in the tough chamise, closer to the lofty timber, is high chaparral, which is also called hard chaparral. High or low—hard, soft, or mixed—all chaparral has in common an always developing, relentlessly intensifying, vital necessity to burst into flame. In a sense, chaparral consumes fire no less than fire consumes chaparral. Fire nourishes and rejuvenates the plants. There are seeds that fall into the soil, stay there indefinitely, and will not germinate except in the aftermath of fire. There are basal buds that sprout only after fire. Droughts are so long, rains so brief, that dead bits of wood and leaves scarcely decay. Instead, they accumulate, thicken, until the plant community is all but strangling in its own duff.

The nutrients in the dead material are being withheld from the soil. When fire comes, it puts the nutrients back in the ground. It clears the terrain for fresh growth. When chaparral has not been burned for thirty years, about half the thicket will be dry dead stuff—twenty-five thousand tons of it in one square mile. The living plants are no less flammable. The chamise, the manzanita—in fact, most chaparral plants—are full of solvent extractives that burn intensely and ignite easily. Their leaves are glossy with oils and resins that seal in moisture during hot dry periods and serve the dual purpose of responding explosively to flame. In the long dry season, and particularly in the fall, air flows southwest toward Los Angeles from the Colorado Plateau and the Basin and Range. Extremely low in moisture, it comes out of the canyon lands and crosses the Mojave Desert. As it drops in altitude, it compresses, becoming even dryer and hotter. It advances in gusts. This is the wind that is sometimes called the foehn. The fire wind. The devil wind. In Los Angeles, it is known as Santa Ana. When chamise and other chaparral plants sense the presence of Santa Ana winds, their level of moisture drops, and they become even more flammable than they were before. The Santa Anas bring what has been described as "instant critical fire weather." Temperatures rise above a hundred degrees. Humidity drops very close to zero. According to Charles Colver, of the United States Forest Service, "moisture evaporates off your eyeballs so fast you have to keep blinking."

Ignitions are for the most part caused by people—through accident or arson. Ten per cent are lightning. Where the Santa Anas collide with local mountain winds, they become so erratic that they can scatter a fire in big flying brands for a long distance in any direction. The frequency and the intensity of the forest fires in the Southern California chaparral are the greatest in the United States, with the possible exception of the wildfires of the New Jersey Pine Barrens. The chaparral fires are considerably more potent than the forest fires Wade Wells saw when he was an undergraduate at the University of Idaho or when he worked as a firefighter in the Pacific Northwest. "Fires in the Pacific Northwest are nothing compared with these chaparral fires," he remarked. "Chaparral fires are almost vicious by comparison. They're so intense. Chaparral is one of the most flammable vegetation complexes there are."

It burns as if it were soaked with gasoline. Chaparral plants typically have multiple stems emerging from a single root crown, and this contributes not only to the density of the thickets but, ultimately, to the surface area of combustible material that stands prepared for flame. Hundreds of acres can be burned clean in minutes. In thick black smoke there is wild orange flame, rising through the canyons like explosion crowns. The canyons serve as chimneys, and in minutes whole mountains are aflame, resembling volcanoes, emitting high columns of fire and smoke. The smoke can rise twenty thousand feet. A force of two thousand people may fight the fire, plus dozens of machines, including squadrons in the air. But Santa Ana firestorms are so violent that they are really beyond all effort at control. From the edge of the city upward, sixteen miles of mountain front have burned to the ridgeline in a single day.

So momentous are these conflagrations that they are long remembered by name: the Canyon Inn Fire, August, 1968, nineteen thousand acres above Arby's by Foothill Boulevard, above the world's foremost container nursery, above the chief executive officer of Mackinac Island Fudge; the Village Fire and the Mill Fire, November, 1975, sixty-five thousand acres above Sunland, Tujunga, La Crescenta, La Cañada. The Mill

Fire, in the words of a foreman at Flood, "burnt the whole front face off."

It is not a great rarity to pick up the *Los Angeles Times* and see a headline like this one, from September 27, 1970:

14 MAJOR FIRES RAGE OUT OF CONTROL
256 HOMES DESTROYED AS
FLAMES BURN 180,000 ACRES

In millennia before Los Angeles settled its plain, the chaparral burned every thirty years or so, as the chaparral does now. The burns of prehistory, in their natural mosaic, were smaller than the ones today. With cleared fire lanes, chemical retardants, and other means of suppressing what is not beyond control, people have conserved fuel in large acreages. When the inevitable fires come, they burn hotter, higher, faster than they ever did in a state of unhindered nature. When the fires end, there is nothing much left on the mountainsides but a thin blanket of ash. The burns are vast and bare. On the sheer declivities where the surface soils were held by chaparral, there is no chaparral.

Fine material tumbles downslope and collects in the waterless beds of streams. It forms large and bulky cones there, to some extent filling the canyons. Under green chaparral, the gravitational movement of bits of soil, particles of sand, and other loose debris goes on month after month, year after year, especially in oversteepened environments, where it can represent more than half of all erosion. After a burn, though, it increases exponentially. It may increase twentyfold, fortyfold, even sixtyfold. This steady tumbling descent of unconsolidated mountain crumbs is known as dry ravel. After a burn, so much dry ravel and other debris becomes piled up and ready to go that to live under one of those canyons is (as many have said) to look up the barrel of a gun.

One would imagine that the first rain would set the whole thing off, but it doesn't. The early-winter rains—and sometimes the rains of a whole season—are not enough to make the great bulk move. Actually, they add to it.

If you walk in a rainstorm on a freshly burned chaparral slope, you notice as you step on the wet ground that the tracks you are making are prints of dry dust. In the course of a conflagration, chaparral soil, which is not much for soaking up water in the first place, experiences a chemical change and, a little below its surface, becomes waterproof. In a Forest Service building at the foot of the mountains Wade Wells keeps some petri dishes and soil samples in order to demonstrate this phenomenon to passing unbelievers. In one dish he puts unburned chaparral soil. It is golden brown. He drips water on it from an eyedropper. The water beads up, stands there for a while, then collapses and spreads into the soil. Why the water hesitates is not well understood but is a great deal more credible than what happens next. Wells fills a dish with a dark soil from burned chaparral. He fills the eyedropper and empties it onto the soil. The water stands up in one large dome. Five minutes later, the dome is still there. Ten minutes later, the dome is still there. Sparkling, tumescent, mycophane, the big bead of water just stands there indefinitely, on top of the impermeable soil. Further demonstrating how waterproof this burned soil really is, Wells pours half a pound of it, like loose brown sugar, into a beaker of water. The soil instantly forms a homuncular blob—integral, immiscible—suspended in the water.

In the slow progression of normal decay, chaparral litter seems to give up to the soil what have been vaguely described as "waxlike complexes of long-chain aliphatic hydrocarbons." These waxy substances are what make unburned chaparral soil somewhat resistant to water, or "slightly nonwettable," as Wells and his colleagues are wont to describe it. When the wildfires burn, and temperatures at the surface of the ground are six or seven hundred centigrade degrees, the soil is so effective as an insulator that the temperature one centimetre below the surface may not be hot enough to boil water. The heavy waxlike substances vaporize at the surface and recondense in the cooler temperatures below. Acting like oil, they coat soil particles and establish the hydrophobic layer—one to six centimetres down. Above that layer, where the waxlike substances are gone, the veneer of burned soil is "wettable." When Wells drips water on a dishful of that, the water soaks in as if the dish were full of Kleenex. When rain falls on burned and denuded ground, it soaks the very thin upper layer but can penetrate no farther. Hiking boots strike hard enough to break through into the dust, but the rain is repelled and goes down the slope. Of all the assembling factors that eventually send debris flows rumbling down the canyons, none is more detonative than the waterproof soil.

In the first rains after a fire, water quickly saturates the thin permeable layer, and liquefied soil drips downhill like runs of excess paint. These miniature debris flows stripe the mountainsides with miniature streambeds—countless scarlike rills that are soon the predominant characteristic of the burned terrain. As more rain comes, each rill is going to deliver a little more debris to the accumulating load in the canyon below. But, more to the point, each rill—its natural levees framing its impermeable bed—will increase the speed of the surface water. As rain sheds off a mountainside like water off a tin roof, the rill network, as it is called, may actually triple the speed, and therefore greatly enhance the power of the runoff. The transport capacity of the watershed—how much bulk it can move—may increase a thousandfold. The rill network is prepared to deliver water with enough force and volume to mobilize the deposits lying in the canyons below. With the appearance of the rills, almost all prerequisites have now sequentially occurred. The muzzle-loader is charged. For a full-scale flat-out debris flow to burst forth from the mountains, the final requirement is a special-intensity storm.

Some of the most concentrated rainfall in the history of the United States has occurred in the San Gabriel Mountains. The oddity of this is about as intense as the rain. Months—seasons—go by in Los Angeles without a fallen drop. Los Angeles is one of the least-rained-upon places in the Western Hemisphere. The mountains are so dry they hum. Erosion by dry ravel greatly exceeds erosion by water. The celebrated Mediterranean climate of Los Angeles owes itself to aridity. While Seattle is receiving its average rainfall of thirty-nine inches a year, Chicago thirty-three, the District of Columbia thirty-nine, and New York City forty-four, Los Angeles is doing well if it gets fifteen. In one year out of every four over the past century, rainfall in Los Angeles has been under ten inches, and once or twice it was around five. That is pure Gobi. When certain storm systems approach Los Angeles, though—storms that come in on a very long reach from far out in the Pacific—they will pick up huge quantities of water from the ocean and just pump it into the mountains. These are by no means annual events, but when they occur they will stir even hydrologists to bandy the name of Noah. In January, 1969, for example, more rain than New York City sees in a year fell in the San Gabriels in nine days. In January, 1943, twenty-six inches fell in twenty-four hours. In February, 1978, just before the Genofiles' house filled with debris, nearly an inch and a half of rain fell in twenty-five minutes. On April 5, 1926, a rain gauge in the San Gabriels collected one inch in one minute.

The really big events result from two, three, four, five storms in a row coming in off the Pacific. In 1980, there were six storms in nine days. Mystically, unnervingly, the heaviest downpours always occur on the watersheds most recently burned. Why this is so is a question that has not been answered. Meteorologists and hydrologists speculate about ash-particle nuclei and heat reflection, but they don't know. The storm cells are extremely compact, deluging typically about ten miles by ten. One inch of rain on a patch that size is seven million two hundred and thirty-two thousand tons of water. In most years, in most places, a winter rain will actually stabilize a mountainside. The water's surface tension helps to hold the slope together. Where there is antecedent fire, water that would otherwise become a binding force hits the rill network, caroms off the soil's waterproof layer, and rides the steep slopes in cataracts into the nearest canyon. It is now a lubricant, its binding properties repelled, its volume concentrating into great hydraulic power. The vintage years present themselves when at least five days of rain put seven inches on the country and immediately thereafter comes the heaviest rainfall of the series. That is when the flint hits the steel, when the sparks fly into the flashpan. On that day, the debris mobilizes.

FIVE MILES INTO the mountains from the edge of the city is a small, obscure, steep-sided watershed of twenty-five hundred acres which is drained by the Middle Fork of Mill Creek, a tributary of the Big Tujunga. The place is so still you can hear the dry ravel. From time to time, you hear the dry cough of semi-automatic weapons. It is the sound of city folk pursuing a hobby. Recreational marksmanship is permitted on the Middle Fork. There are eight million people just down the wash, and they shoot some interesting guns. Amos Lewis, who covered the region as a deputy sheriff for twenty-five years, once found beside the Angeles Crest Highway "a gun you could hide behind your tie—you'd think it was a tie clip." He has also seen enough muzzle-loaders to have made a difference in the Battle of Long Island. In an imaginative, life-loving city, there will always be people with a need to fire antique weapons. On July 24, 1977, a marksman on the Middle Fork rammed Kleenex down his barrel instead of cloth wadding. Under the Kleenex was black powder. In black powder there is more of an incendiary risk than there is in the smokeless kind. When the rifle fired, flaming Kleenex shot out the muzzle and burned down three thousand eight hundred and sixty acres, including the entire watershed of the Middle Fork.

It was a textbook situation—a bowl in the mountains filled with hard chaparral that had not been touched by fire in ninety-nine years. The older chaparral becomes, the hotter it burns. In its first ten years of new growth, it is all but incombustible. After twenty years, its renewed flammability curves sharply upward. It burns, usually, before it is forty years old. The hotter the fire, the more likely a debris flow—and the greater the volume when it comes. The century-old fuel of the Middle Fork was so combustible that afterward there were not even stumps. The slopes looked sandpapered. The streambed, already loaded, piled even higher with dry ravel. The Middle Fire, as the burn was known, was cause for particular alarm, because a small settlement was a mile downstream. Its name— Hidden Springs—contained more prophecy than its residents seemed prepared to imagine. Three hundred and ninety thousand cubic yards of loose debris was gathered just above them, awaiting mobilization.

Dan Davis and Hadi Norouzi, L.A.C.F.C.D. engineers, went up there after the burn to tell the people what they might expect. In midsummer, it is not a simple matter to envision a winter flood if you are leaning on a boulder by a desiccated creek. "We spent a lot of time trying to prevent a disaster from occurring," Davis said recently. "The fact that people would not believe what *could* happen was disappointing, actually. We held meetings. We said, 'There's nothing we can do for you. Telephones are going to go out. Mud will close the road. You're abandoned. If you're here, get to high ground.' " There was no debris basin, of course. This was a hamlet in the mountains, not a subdivision at the front. Conditions were elemental and pristine. "We walked people through escape routes," he went on. "We told them the story of fire and rain. We said, 'If heavy rain starts, you've got fifteen to thirty minutes to get out.' "

Norouzi told them they were so heavily threatened that no amount of sandbags, barricades, or deflection walls was ever going to help them. "There is nothing you can build that will protect you."

Half a year went by, and nothing stirred. Cal Drake went on making jewelry in his streamside apartment. He and his wife, Mary, shared a one-story triplex with two other couples. The Drakes, from the city, had moved to Hidden Springs two years before, in quest of a "quiet life." Elva Lewis, wife of Amos the sheriff, went on running her roadside café. Gabe Hinterberg stayed open for business at the Hidden Springs Lodge. In December and January, there was an unusual amount of rain, but no flood. By the end of the first week of

February, there had been eighteen inches in all. Then, in the next three days, came enough additional rain to make this the winter of the greatest rainfall of the twentieth century, exceeded only by 1884 and 1890 in the records of Los Angeles County. The National Oceanic and Atmospheric Administration selected the word "monstrous" to befit the culminating February storm, in which almost a foot of rain fell in twenty-four hours, and, in the greatest all-out burst, an inch and a half in five minutes. This was the storm that sent the debris down Pine Cone Road, overtopped the Zachau Basin, mobilized the corpses in the Verdugo Hills. In the small valley of the Middle Fork, upon the scorched impenetrable ground, three million tons of water fell in one day.

Toward midnight February 9, an accidental fire broke out in a small building of Gabe Hinterberg's. A fire truck eventually came. Half a dozen people fought the fire, assisted by the heavy rain. One of them was George Scribner. The five-minute spike of greatest downpour occurred at about one-thirty. Half an hour later, George said, "Hey, we got the fire put out."

Gabe said, "Good deal."

And then Gabe and George were dead.

Amos Lewis, nearby, was holding a fire hose in his hand and was attempting to prevent it from kinking. In his concentration, he did not see danger coming. He heard nothing ominous. He only felt the hose draw taut. Through his peripheral vision he became aware that the fire truck—with the hose connected to it—was somehow moving sideways. Seconds later, Amos Lewis, too, was swept away.

The snout of the debris flow was twenty feet high, tapering behind. Debris flows sometimes ooze along, and sometimes move as fast as the fastest river rapids. The huge dark snout was moving nearly five hundred feet a minute and the rest of the flow behind was coming twice as fast, making roll waves as it piled forward against itself—this great slug, as geologists would describe it, this discrete slug, this heaving violence of wet cement. Already included in the debris were propane tanks, outbuildings, picnic tables, canyon live oaks, alders, sycamores, cottonwoods, a Lincoln Continental, an Oldsmobile, and countless boulders five feet thick. All this was spread wide a couple of hundred feet, and as the debris flow went through Hidden Springs it tore out more trees, picked up house trailers and more cars and more boulders, and knocked Gabe Hinterberg's lodge completely off its foundation. Mary and Cal Drake were standing in their living room when a wall came off. "We got outside somehow," he said later. "I just got away. She was trying to follow me. Evidently, her feet slipped out from under her. She slid right down into the main channel." The family next door were picked up and pushed against their own ceiling. Two were carried away. Whole houses were torn loose with people inside them. A house was ripped in half. A bridge was obliterated. A large part of town was carried a mile downstream and buried in the reservoir behind Big Tujunga Dam. Thirteen people were part of the debris. Most of the bodies were never found.

As Amos Lewis suddenly found himself struggling in the viscous flow, he more or less bumped into a whirling pickup coming down in the debris from who knows where upstream. One of the roll waves picked him up and threw him into the back of the truck. As the vehicle spun around and around, it neared one bank. Lewis saw an overhanging limb. He reached for it, caught it, and pulled himself above the rocky flow. Years later, just about where this had happened, he told Wade Wells and me the story. "I got pushed to one side," he said as he finished. "I lucked out." Lewis is a prematurely white-haired man with a white beard and dark-brown eyes. On this day in late spring, his muscular build and deeply tanned skin were amply displayed by a general absence of clothing. He wore bluejean shorts, white socks, mountain boots, and nothing else. When people began to discover human remains in the reservoir, he had gone in his patrol car to investigate the fate of his neighbors. "I had to go roll on them calls," he said. "A deputy sheriff has to roll on any type of body being found. I carried out at least four, maybe five, skulls."

The thirteen people who died in Hidden Springs were roughly a third of the year-round community; there was a much larger summer population. The main house of Lutherglen, a resort-retreat of the First English Evangelical Lutheran Church, remained standing but in ruins. Houses that stayed put were gouged out like peppers and stuffed with rocks. Lewis gestured across the canyon—across foundations with no houses on them, bolts sticking up out of cinder blocks where sills had been ripped away—toward some skeletal frames made of two-by-fours. "They used to be trailer stalls," he said. "The people left their cars by the river and walked up the bank to the trailers. The cars ended up in the dam." The First English Evangelical Lutherans sued the Los Angeles County Flood Control District for twenty million dollars. The judge threw the case out of court—followed, moments later, by the collection plate. Since the act in question was God's, the defendant might as well have been the plaintiff, and the Plaintiff the target of the suit.

I remarked to Lewis, who is now retired as sheriff, that I thought I'd heard a machine gun earlier in the day. "I worked the canyon car here for twenty-five years," he said. "I probably rolled on a minimum of a hundred and fifty calls where people said they heard machine guns. I never saw a machine gun."

Wells was attentive to this remark, raising his eyes with interest. Behind his mild ecological look—his tortoise-shell glasses, his amiable scientific manner—lay a colonel's affection for ordnance. At the time, in the Reserve, he was a lieutenant colonel and rising. He'd been on active duty seven years, two in Vietnam. He told me one day that if California were to secede from the United States it would be one of the richest countries in the world and, with its present units of the National Guard, be among the best defended. "You can take a file and in fifteen minutes make an automatic weapon out of an M1," he said to Amos Lewis. "It can sound like a machine gun."

This set off a long and highly technical discussion between the scholarly hydrologist and the shirtless mountaineer, each slipping into a second self against a backdrop of huge boulders that had been somewhere else a short time before and had been delivered by a force that was high in the kiloton range. Most of the mud, sand, and rock had gone into the Big Tujunga, behind the dam, and the county had spent more than two million dollars taking it out. The debris that had stayed in the valley closely resembled glacial debris—chaotic, unsorted till, a round-rock mélange. Far up the hillsides framing the valley, some of it clung like bits of plaster stuck to an old wall, thus recording the high edges of the discrete slug, where six hundred thousand tons went by.

WHEN YOU WALK in the stream valleys of the San Gabriels, you will see rocks the size of heads wedged among the branches of trees. In a small tight valley called Trail Canyon, I saw two boulders that were a good deal wider than the bed of the brook that had carried and rounded them. They were bigger than school buses. Surrounded by lesser debris, they had moved a long distance in its company. At a guess—from their dimensions and specific gravity—the aggregate weight of the two rocks was a hundred and sixty tons.

In February, 1978, a boulder weighing three hundred and fifteen tons ended up on a residential street about a third of a mile inside the Los Angeles city limits. Through some neighborhoods, boulders in great numbers advance like Chinese checkers. People pile them up against fences, use them in retaining walls. When Dan Davis was working for Flood, he found debris—on an urban thoroughfare after a storm—a mile and a half from the nearest debris basin. ("When I saw that, I knew we had a real problem.") In 1938, a restaurant on the main street of Sierra Madre was destroyed by invading boulders. Two-foot boulders rumbled through Claremont, coming to a stop three miles from the mountain front. Five miles from the front you can see boulders a foot in diameter. If you ask people how the rocks got there, they assume it was by a process that is no longer functioning. If you suggest that the rocks may have come from the mountains, people say, "No way." Off the eastern end of the San Gabriels, rocks the size of soccer balls are eight miles south of the front.

Building stones in places like Glendora and Covina were delivered by streams from high in the mountains. The stream-rounded rock is more vulnerable to earthquake than bricks would be, but bricks are not shipped F.O.B. by God, and in a land of kaleidoscopic risks what is one more if the rocks are free? Mike Rubel's castle, in Glendora, is made of stream-rounded debris in sizes approximating cannonballs. Dunsinane was not much larger than this suburban home. The ground level of Rubel's castle is twenty-two thousand square feet. From its battlements rise towers sixty-seven feet high and seventy-four feet high, built with San Gabriel boulders and store-bought cement. There are six towers, four set in the walls and two in the courtyard freestanding. Bees live in the Bee Tower, and emerge through archery slits. All around the walls, muzzles of cannons protrude from crenels that are lined with shark-fin glass.

The intensity of the electronic surveillance is high, but the owner is not unfriendly. He likes to sit on a balcony above the courtyard, looking out over his walls and through the crowns of palms at the ridgeline of the mountains. He is a large man to the point of private tailoring. He began his castle in 1959 and completed it in 1985. When he had been working on the project ten years, he took an unexpected delivery of building materials in the form of a debris slug that breached his defenses, untimbered his portcullises, and got into the inner bailey.

"The ground was shaking just like an earthquake. In the washes, the water was going three billion miles an hour. You could hear the boulders rumbling. It was marvellous."

As a result, there is now a twelve-foot curtain wall on the periphery of the castle. Rubel calls his domain, which is surrounded by commoner houses on a most conventional street, the Kingdom of Rubelia. Numerous crafts are practiced there, and he has a hand-set-printing operation called the Pharm Press. In the Kingdom of Rubelia, F is Ph and Ph is F. There are hand-cranked phorges in the blacksmith phoundry. There are potters' wheels, looms, and lathes.

Sitting beside him on his balcony and dreamily looking at the mountain peaks, I said, "The castle is obviously the result of something."

Rubel said, "Yes. A genetic defect."

Rubel explained that he had built the castle with the help of numerous friends—friends from his days in Citrus High School, friends from his briefer days at Cal Poly. "We were twenty-year-old kids," he said. "And we were flunking out of school. We said, 'If we can't amount to anything, we might as well build a castle.' "

Prince Philip of Great Britain, who is not a Rubelian and gets no F, has made two visits to Rubel's castle.

Cal Poly—the California State Polytechnic University—is not to be confused with Caltech. I bring this up because I went to Caltech one day and, in a very impromptu manner, asked to see a geologist. Any geologist. It had not been my purpose, in pursuing the present theme, to get into the deep geology. I meant to roam the mountains and the mountain front with foresters and engineers, to talk to people living on the urban edge, to interview people who sell the edge—a foreign correspondent covering the battle from behind both lines. But not beneath them. This was a planned vacation from projects in geology—the continuation of a holiday that had begun with stream capture in the lower Mississippi and had spread forth into such innocent milieus as eruptions in Iceland and flowing red lava in Hawaii. Now, in Los Angeles, I had been avoiding geologists in the way that one tries to avoid visits to medical doctors. All had gone well for a matter of weeks, but then, one morning, I just happened to be in Pasadena looking up into the veiled chimeric mountains, and severe symptoms began to develop. Right off the street—in much the way that a needful patient would seek out a Doc-in-the-Box—I walked into the geology department of the California Institute of Technology, found the departmental office, and asked for professional help.

After a short wait, spent leafing through a magazine, I

was shown into the office of Leon Silver, whom I knew only by reputation—an isotope geologist whose exacting contributions to geochronology have not repressed his interest in crustal settings, global tectonics, the Big Picture. An ebullient man, husky, in his sixties, he spread out the local sheets from the geologic map of California for a brief rehearsal of the rocks and faults before leading me to the roof of the building, where he continued his diagnosis in the panoramic presence of the rock itself. The roof was flat, a deck. Funnel vents and other apparatus gave the impression that the Caltech geology department was a cruise ship in the lee of seventy miles of mountains.

The institution as a whole, in its remarkable beauty and surprisingly compact size, is sort of a bonsai university—with pools, rialtos, inclined gardens—above which the mountains seem all the more immense. Silver said that if I was looking for first causes in the matter that concerned me I had come to the right place. "The geology provides the debris," he went on. "The San Gabes are a climber's nightmare. Several people a year die on the incompetent rock."

"Yes," I said. "The rock up there is really rotten."

Silver seemed offended. Drawing himself up, he said, "I beg your pardon, sir. It is not rotten. It is shattered." The region was a tracery of faults, like cracks in ancient paint. The mountains were divided by faults, defined by faults, and framed by them as well: on the near side, the Raymond Fault, the Sierra Madre Fault, the Cucamonga Fault; on the far side, the San Andreas Fault. The rock of the San Gabriels had been battered and broken by the earthquakes on these and related faults. In 1971, Silver had flown over the San Gabes immediately after an earthquake that reached 6.2 on the Richter scale. Like artillery shells randomly exploding, the aftershocks were sending up dust in puffs all over the landscape. Something like that would add quite a bit, he said, "to the debris potential." Some of the rock up there had become so unstable that whole hunks of the terrain were moving like glaciers. One mountaintop was heading south like a cap tipping down on a forehead. Things like that had been going on for so long that the mountains were in many places loaded with debris from ancient landslides—prime material, prepared to flow. "The ultimate origin of the debris flows," he said, "is the continuous tectonic front that has made this one of the steepest mountain fronts in North America and produced a wilderness situation not a hundred metres from people's houses."

The continuous tectonic front is where the North American and Pacific Plates are sliding past each other—where Bakersfield moves toward Mexico City while Burbank heads for Alaska. Between Bakersfield and Burbank lie the San Gabriel Mountains. With the San Bernardino Mountains east of them, they trend east-west, forming a kink in the coastal ranges that come down from San Francisco and go on to Baja California. The kink conforms to a bend in the San Andreas Fault, which runs along the inland base of the mountains. The kink looks like this:

It could be a tiptoeing h. It resembles a prize-winning chair. Los Angeles is like a wad of gum stuck to the bottom of the chair. The mountains are one continuous system, but its segments are variously named. The upper stretch is called the Coast Ranges. The lower leg is called the Peninsular Ranges. The kink is called the Transverse Ranges.

My hieroglyph represents, of course, not only the mountains but the flanking San Andreas Fault, which comes up from the Gulf of California, bends left around Los Angeles, then goes on to San Francisco and north below the sea. As if this regional context were not large enough, Silver now placed it in a larger one. The East Pacific Rise, the ocean-basin spreading center away from which the Pacific Plate and other plates are moving, sinuously makes its way from the latitude of Tierra del Fuego all the way north to Mexico, where it enters the Gulf of California. The East Pacific Rise has splintered Mexico and carried Baja California away from the mainland—much as the Carlsberg Ridge has cracked open the deserts of Afro-Arabia and made the Red Sea. Baja is not moving due west, as one might guess from a glance at a map, but north by northwest, with the rest of the Pacific Plate. The cumulative power of this northward motion presses on the kink in the San Andreas, helping the mountains rise.

That much has long seemed obvious: as the two sides of the San Andreas slide by each other, they compress the landscape at the kink. It has been considerably less obvious that a compressional force accompanies the great fault wherever it goes. In the past, the building of the Coast Ranges and the Peninsular Ranges was in no way attributed to the San Andreas Fault. A paper published in *Science* in November, 1987—and signed by enough geologists to make a quorum at the Rose Bowl—offers evidence that the San Andreas has folded its flanking country, much as a moving boat crossing calm waters will send off lateral waves. The great compression at the kink is withal the most intense. The Coast Ranges and the Peninsular Ranges are generally smaller than the Transverse Ranges. The San Gabriels are being compressed about a tenth of an inch a year.

Why the kink is there in the first place is "not well understood." Just to the northeast, though, in the Great Basin of Utah and Nevada, the earth's mantle is close, the earth's crust is thin and stretching. In the past few million years, the geographic coordinates of Reno and Salt Lake—at the western and eastern extremes of the Great Basin—have moved apart sixty miles. This large new subdivision of the regional tectonics is in every way as entrancing as it is enigmatic. Almost all of California may be headed out to sea. Already, the east-west stretching of the Great Basin has put Reno west of Los Angeles, and it may be what has bent the San Andreas Fault.

Some of the rock of the San Gabriels is two hundred times as old as the San Andreas Fault, which has been in existence for less than a five-hundredth of the history of the world. Plates come and go—splitting, welding, changing through time, travelling long distances. Before the present North American and Pacific Plates began to work on this particular rock, Silver said, it may have been "bashed around in Mexico twice and perhaps across the Pacific before that." He continued, "It's a bedrock ridge up there. It's a weirdo wonderful block of rocks, the most complicated mountain range in North America. It includes the oldest rocks on the West Coast. The San Gabes look like a flake kicked around on plate boundaries for hundreds of millions of years."

The Santa Monica Mountains, a sort of footnote to the big contiguous ranges, stood off to the southwest of us, discrete and small. Like any number of lesser hills freestanding in the region, they were flexures of the San Andreas system. Oil people had found pay in the traps formed by such flexures. The Santa Monica Mountains were as shattered as the San Gabes. The several debris basins in the Santa Monicas had worked with varying success. People had died in their beds there, buried alive by debris.

The San Gabriels were rising faster than they were disintegrating, Silver said. The debris basins had given geomorphologists an unparalleled opportunity to calculate erosion rates. They could even determine how much mountain is removed by a single storm. On the average, about seven tons disappear from each acre each year—coming off the mountains and heading for town.

Between the geology-department roof and the San Gabriels, the city gradually rose. A very long, ramplike, and remarkably consistent incline ended in the sheerness of the mountain wall. This broad uniform slope is where the seven tons an acre had emerged from the mountains, year upon year for a number of millions of years—accumulating as detrital cones, also known as fans. Broad at the bottom, narrow at the top, the fans were like spilled grain piling up at the edge of a bin. There were so many of them, coming down from stream after stream, that they had long since coalesced, forming a tilted platform, which the Spaniards had called the bajada.

"I used to live on the mountain front," Silver said. "By Devils Gate, at the mouth of Arroyo Seco. We could hear the big knockers go by—the three-metre boulders. The whole front face of the San Gabes is processed."

"Processed?"

"Shattered and broken. It is therefore vulnerable to landsliding, to undercutting by the streams, to acceleration by local earthquakes, to debris flows."

"Why does anybody live there?"

"They're not well informed. Most folks don't know the story of the fire-flood sequence. When it happens in the next canyon, they say, 'Thank God it didn't happen here.' "

"Why would a geologist live there?"

"It's a calculated risk. The higher you build, the cooler

it is. There are great views. And at night, up there, the cool air off the mountains flows down and pushes the dirty air masses back. The head of our seismological laboratory lives on the mountain front. In fact, most of the Caltech geology department lives on the mountain front."

"Where do *you* live?"

"Way out on the fan."

Silver passed me along to his colleague Barclay Kamb— the tectonophysicist, X-ray crystallographer, and glaciologist, who discovered, among other things, the structures of the high-pressure forms of ice: ice II through ice IX. Kamb once studied the Sierra Madre Fault Zone on the San Gabriel mountain front, and walked the relevant canyons. Recently, he has been using a surging glacier near Yakutat as a laboratory for the study of how rocks move, since ice deforms in much the way that rock does. He was about to leave for Alaska when I dropped in on him in his office. His mother was there, his father, and his son Linus, who was named for Kamb's father-in-law, Linus Pauling. In a swirl of ropes, ice axes, grad students, and relatives, Kamb, who has been described by another colleague as "the smartest man in the world," tracked six conversations simultaneously, one of which summarized concisely his sense of flowing debris. "There's a street in Altadena called Boulder," he began. "It is called Boulder for a very good reason. It is subject to severe threat. Boulder Road, below the Rubio Debris Basin, is the former course of Rubio Creek. You see encroachment of human habitation in many areas like that, which are most at risk. Above the debris basins, there are crib structures in the canyons. The theory is to prevent sediment from coming out of the mouths of the canyons. I think most geologists would say that is ridiculous. You're not changing the source of the sediment. You are just storing sediment.

Those cribworks are less strong than nature's own constructs. The idea that you can prevent the sediment from coming out is meddling with the works of nature. Sooner or later, a flood will wipe out those small dams and scatter the debris. Everything you store might come out in one event. We're talking human time—not geologic time." Kamb lives in Pasadena, close by the mountain front.

Just upstairs was Andrew Ingersoll, the planetary scientist. In the San Gabriels, he had lived behind the lines. In the nineteen-sixties, he moved his family into a cabin that was so far up Big Santa Anita Canyon that they had to hike a mile and a quarter just to get to their car. They leased the place from the Forest Service. When they moved in, the children were three and four. Ingersoll was an assistant professor. "My colleagues in the geology department thought I was becoming a permanent hippie," he said. "But in those days everybody was some sort of hippie." The canyon was full of crib structures, arresting debris. Ingersoll did not know how to make sense of them unless they were "an example of bureaucracy doing something for its own sake." (In any case, the small wash above the Ingersolls' cabin was unprotected.) In January of 1969, during a nine-day series of storms, twelve inches of rain fell in one night. A debris flow hit the cabin, broke through a wall, and delivered three feet of mud, innumerable rocks, and one oak to the Ingersolls. The family regarded this as "just a lot of fun," he said, and continued, "Those little dams must have been nearly insignificant. They were based on the experience of Swiss farmers, and this may have been a totally different situation. It might have been a very poor concept to try to control the San Gabriels."

I also met Vito Vanoni, who is now a professor emeritus. A formal, small, wiry man with a husky voice and a sweet smile, he is a civil engineer, and a founding and still central figure in Caltech's Environmental Quality Laboratory. "That's an awful pile of rock and dirt up there, and we're proposing to hold it back," he said. "To do something like that is extremely expensive, but there are so many of us here to pay the bill, to protect those who insist on living up there. Our zoning is not strong enough to prevent this. The forces of development are hard to oppose. Most people who buy property in those areas never see the map and wouldn't know what they were looking at if they saw one. Very few are aware. When they see the concrete stream channels, I don't know what they think. How many people really realize why the channels are there and why they are as big as they are? You can't build a channel without a debris basin, or the debris will fill up the channel and then start sashaying back and forth. Debris basins have been built in response to the need of the community—after people have had sediment in their living rooms."

I asked Vanoni where he lived.

"Up there," he said. "Below Eaton Basin—since 1949. Like my neighbors, I figure that I'm protected. I haven't seen anything across my yard yet." After a pause, he added, "If they should have a failure up there, I'm afraid I'd get wet." There was a longer pause, then another sweet smile, and he said, "I live a hundred yards from the Raymond Fault."

IN GLENDORA, I once came upon a solid block of citrus trees surrounded by residential streets. In all directions from this dark-green stamp sprawled the megalopolitan sprawl—the vast groves of houses, many of them with the pantiled roofs, the pocket arcades, the open-trussed porches, the adobe walls that Reyner Banham called Neo-Churrigueresque and much admired for making "both ancestral and environmental sense." From the block of citrus the houses continued west unremittingly, east and south indefinitely, and north about eight hundred yards, where they were stopped by the mountain front. The land under the houses was almost all under citrus just a few decades ago—from the orchards of Glendora and Covina east to the orchards of Pomona and west to Pasadena. Glendora's memorial citrus was surrounded by a low stone wall of stream-rounded granodiorite, the principal rock of San Gabriel Mountain debris. As if the dark trees were not beauty enough, the largest bougainvillea in the United States adorned an edge of this vestigial scene. The bougainvillea was planted in 1901, and now, in urban isolation in the Neo-Churrigueresque, was commemorated by a plaque as California Registered Historical Landmark No. 912:

> *The parent stock was brought to California*
> *by a whaling ship about 1870, and the vines*
> *survive as one of the best examples remaining*
> *of the early 20th Century promotional image of*
> *California as a paradise.*

The block of citrus has since been destroyed to make space for condominiums.

Just down the street was the office of Charles Colver, who grew up in a family of orange growers and was now manager of the experimental forest in the mountains above. He lived in his family's original ranch house, its land reduced to a third of an acre. He had learned in his sixty-odd years that the "C" of the vitamin stands for "Chekhov." A tall sandy-haired man with a long face and an attractively slow cadence of speech, he had seen many slurries of flowing debris, in each a story to tell: "After midnight, the water was rising, still clear, when, with no warning, a big bulk load came through there—a huge wall of mud with a lot of trees in it. It overtopped the flume. The debris was eight feet high."

When the big bulk loads came out of the canyons and went into the orchards of citrus, they knocked over some trees. The debris stayed where it was, and the growers planted new trees. This was the process that had built the fans, and was still building the fans. In a way that housing never could, the citrus accommodated, even when the debris was sixteen feet deep. "We have encroached on nature," Colver remarked to me one day. "When we diddle with Mother Nature, we mess up things. We're living on a floodplain. To look at it, you'd think it was flat, but there's nine hundred feet of difference from Glendora to the ocean. The alluvial fans are that deep. The types of flows that built them go on trying to build them, where we are trying to live."

Before the citrus, there were ranches; before the ranches, Indians; before the Indians, the primeval scene: huge unencumbered alluvial fans leaning into the fast-rising mountains beside the hazy plain—the broadest coastal lowland in all of California. Despite the sea fog and the general lack of more concentrated water, the climate was so congenial that when people discovered it they were willing to struggle to live in it. The people of Asuksangna were among the first. Their village has become Azusa. For agricultural purposes, they burned grasslands in spring. Inadvertently, they torched untold acreages of chaparral and set up innumerable debris flows. Lightning ignited what the Indians missed. Easily, they adapted to the consequent flows. Ranchers, when they came, settled by the canyons to guard the trickling water, and shot one another over rights to the water, which occasionally carried loads big enough to kill them all. Huge contiguous ranches lined the San Gabriel front—Rancho San José, Rancho Azusa de Duarte, Rancho Santa Anita, Rancho La Cañada, Rancho Tujunga, Rancho Ex-Mission de San Fernando. During the long drought of the eighteen-sixties, horses and mules went into the mountains to bring out water in tanks. The effort notwithstanding, the bones of parched cattle were scattered about the plain. The message was clear: this environment was no less hostile than it was appealing. When the railroads arrived (1876, 1885), they set off a population splurge that has since become known as California's second gold rush. The people rushing in were farmers, and the gold was oranges. From Kansas City, the fare to Los Angeles was for a brief time a dollar. The citrus orchards were established in small units, Colver said, "ten to fifteen acres each—all that one man could tend to." There were soon a hundred thousand acres of citrus. Ditches from the mountains irrigated the trees. Once in a while, the ditches filled with debris. In 1884, in the aftermath of a fire in Soledad Canyon, debris tore away the tracks of the Southern Pacific. (In 1978, in the aftermath of fire, debris wrecked the tracks in the same place.)

In the eighteen-seventies, to connect agricultural towns, local railways had begun to climb the bajada. Long straight avenues are there now, steadily rising three and four miles. At least one railway was mule-drawn. When the mules made it to the top of the fan, they went around to the back of the train and got onto a special flatcar, on which they rode downhill. In the eighteen-nineties, electricity replaced the mules, and the street railways began to assume almost the exact pattern of the freeways that have replaced them. Under the influence of the Pacific Electric Railway, communities began to coalesce, like the alluvial fans.

235)

For many decades, there was a moat of oranges between the built-up metropolis and the mountain front. After the wildfires of 1913 resulted in the flows of 1914, inspiring, in 1915, the creation of the Los Angeles County Flood Control District, debris was not thought of as the essence of the threat. The operating word was "Flood."

Just after the beginnings of Flood Control, the naturalist John Burroughs lived for a time in Pasadena Glen. The house had broad window seats, wooden casements, a big fireplace made of stream-rounded boulders. Cabinlike in atmosphere, it suggested something a good deal more remote than a suburb, as well it might, for in its steep and shaded canyon it was truly in the wild and beyond the march of towns. This one-street neighborhood, fingering into the mountains under evergreen oaks, still has an escapist look, although it has all but been caught by the city. Taking in the whole neighborhood in a single remark, Chuck Colver will tell you, "Those people are crazy." He means that Pasadena Glen has the compact dimensions of a bobsled run, and the disassembling mountains hover above it. He means that from time to time all hell will break forth from the mountains. To the question "Why, then, do people live there?" the answer seems to be that they are like John Burroughs: they would rather defy nature than live without it. The glen is so narrow that its houses are perched between streambanks and canyon walls. The nearest debris basin is below them, and therefore not meant to help them. The narrow roadway—crossing, recrossing—laces the stream. In contexts in which most people would say "my driveway," people in the glen often say "my bridge." To get to John Burroughs' place you go over a short private ford. That is the entire driveway. A family named Horton lives where John Burroughs lived. The Hortons have watched flowing debris demolish their neighbors' houses. "People come in and live—as we do—where we really shouldn't live," Mel Horton remarked to me, tendering the explanation that periods between serious floods are often long.

"It's a fantastic place to be in a storm," his wife, Barbara, said. "You hear a sound like giant castanets—boulders clicking together. They're not pebbles. And there is a scent, which is absolutely heavenly, of the crushed chaparral plants. It's so fragrant and beautiful it's eerie to have it associated with something so terrifying. And, God knows, it is terrifying."

"So why do you live here?"

"Freedom," Mel said. "You go to the city for a living. For provisions. Here at home, you don't have a sense of the city. You feel renewed. There is no sense of encroachment. Although there are fifty-five homes in this canyon, the open mountains begin at this point. You're in the sycamores, the oaks. You feel removed from a sense of being surrounded."

Barbara said, "City people have trouble with the stillness here in this canyon. But something is always going on here, among the ground squirrels, the wrens, the wood pewees. There was a butterfly that chased off a bird." A rooster crowed. Barbara continued, "One needs some sense of perspective if you live in a place that has built-in hazards. But, then, what place doesn't? You can have an earthquake anywhere."

Their daughter, Alison, who lives in Alaska and was about to return there, said, "In Anchorage, I'm in a more urban environment than I ever lived in in Los Angeles. When I was in junior high school and went to friends' houses outside the glen, I felt hemmed in. Home was the escape for me here. In Alaska, I escape from home."

The Hortons' friend Roland Case Ross, emeritus professor of biology at California State University, happened to be there as well, and he spoke up to say that the Hortons and everyone else in Pasadena Glen were "recluses—seekers of solace, nature contact, and purism." As if Ross were not. Ross remembered the mountains when the state animal was not yet extinct in the state. The state animal is the grizzly bear. Ross followed bear trails through "fog-drip country" in "chaparral litter to your knees." He knew the wild mountainsides, and how to get back to "Loss Anglus." One rainy day, he was invited to a restaurant in Altadena, and he firmly declined. "Only a damned fool would put a restaurant in a floodplain," he said. The restaurant was destroyed that evening by a debris flow.

Mel Horton said that Pasadena Glen, like so many other canyons where houses penetrate the mountain front, maintains its serenity for years on end until "four inches of rain falls in one hour."

Barbara said, "It catches your attention."

In 1969, after less than two inches of rain fell in one hour the Hortons' house was shaking, debris gouged the streambed and quadrupled its width, neighbors' houses were undermined to the point of destruction, telephone poles danced about like sticks, a big tree trunk heaved into the air and shorted the power line, and debris that missed the Hortons' house went through the house below them. Recalling that night, Barbara Horton said, "Those boulders that came down were like pianos. They formed a dam fifty feet above our bridge. Then the water broke around the sides. The larger part went across the road, toward the Childresses'. Childress wanted to run out his back door with his children. He couldn't. The flow went around his house and destroyed the Hemmingers' place. Hemminger was on the roof rescuing his cat. The Hemmingers eventually moved to another house in the glen."

After the 1969 event, the Forest Service, with "the blessing of the Flood Control people," lined the stream with Gunite in Pasadena Glen. I walked upstream with the Hortons, to the head of the populated canyon, to see what the Forest Service people had done. They had changed a natural stream into a concrete trough. Barbara Horton said, "It's something everybody tries to forget." The Gunite banks were topped on both sides with hundreds of large granitic boulders. Protecting the uppermost property in the glen was a twelve-foot stone barricade. In the stillness, a sound came from the head of the canyon and amplified as we drew near. It was the sound of falling water—falling off a ledge and three stories down, between narrow chasm walls. There were trees rooted in the chasm walls, and brown grass on a slope above the falls. There had been no rain in four months. The sound of the water was as unexpected as the sound of falling water on the moon.

BY THE NINETEEN-THIRTIES, the Thoreauvian recluses were well established in the mouths of canyons, and the city had not yet approached them. Looking down the fan across the orange orchards, they could call themselves remote. It is coolness that makes an orange orange. The air that fell in the evening off the San Gabriel front touched the dark trees with color, and the towns among the foothills were still discrete.

In November, 1933, the chaparral burned in numerous watersheds above Pasadena, La Cañada, La Crescenta, and Montrose, and slopes were left black and bare. Rainfall in amounts that the Flood Control District called a "Noah-type storm" followed in the last days of the year, mobilizing, on January 1, 1934, a number of almost simultaneous debris flows that came out of the mountains, went through the orchards and into the towns, killed dozens of people, destroyed hundreds of houses, and left boulders the size of icebergs far down the fans. (In the middle of all this chaos, the football team of Columbia University went into the Arroyo Seco and defeated Stanford in the Rose Bowl.) Out of Pickens Canyon came a debris slug of such magnitude that it travelled all the way to Foothill Boulevard, crossed it, and passed through the business district of Montrose. A boulder eight feet in diameter came to rest on the main street of town, three miles south of the mountain front. This was the same Montrose to which Jackie Gen- ofile would one day retreat in order to feel safe. The New Year's Day Flood, as people still refer to it, killed thirty-four in Montrose and neighboring towns, and ruined nearly five hundred houses. All over the bajada, Model A's were so deeply buried that their square roofs stuck out of the mud like rafts. Streets of La Crescenta, a mile downhill from where the Gen- ofiles live now, were like the braided rivers of Alaska, with channels of water looping past islands of debris.

The working gravel pit that filled up and fortuitously became the prototype debris basin was in Haines Canyon, just above the village of Tujunga, whose civic infrastructure it happily preserved from otherwise certain annihilation. While the mud, sand, and boulders that were deeply spread on the alluvial fans were demonstrating that a flood of rock was a far greater menace than a flood of water, this inspirational gravel pit was showing what to do about it. With enough money— enough steam shovels, enough dump trucks, enough basins cupped beneath the falling hills—Los Angeles could defy the mountains, and append to an already impressive list one more flout in the face of nature.

After Pearl Harbor, rapidly expanding war industries drew people to Los Angeles from all over the United States. They worked mainly in the oil, steel, and aircraft industries and in plastics factories that made butadiene as a substitute for rubber. The ascending effluents of the smelters, refineries, mills, and factories added a great burden to the marine fog layer—made heavier still as the work force moved about in cars. To describe this ochre cumulus, the world's shortest portmanteau word, which had been coined around 1905, was borrowed from London.

Housing developments had to be created. The land under the orchards was an obvious choice, and a lot of citrus fell in the war. Like a spider plant or a wandering Jew, an orange tree is psychologically sensitive. Not surprisingly, a virus broke out in the orchards—a strain known to pomologists as "the quick decline."

When the war ended, the quick decline did not. As Chuck Colver well remembers, "This was a mortal blow. If a grower didn't suffer from it, he thought he would, so he sold to a developer as an easy way out. After the war, the new people stayed on, and it was readily apparent that agriculture and urbanization were not compatible. Citrus relied on smudge pots to protect the trees against freezes. It used fertilizer that was smelly. It created dust in the tilling. It bred flies. Big slow trucks went around full of oranges. Everybody had tolerated all this when oranges were the sole economy, but it was a nuisance to the newcomers. They threw garbage in the orchards. They stole oranges. And, above all, they complained. They passed laws against smelly fertilizer, against smudge smoke, against pesticides. Citrus could not compete. Water became too high-priced. Smog began to affect the trees. The size and quality of fruit deteriorated. A superior product became an inferior product."

Fifty miles of citrus disappeared as the communities coalesced and the megalopolitan census passed eight million. Los Angeles did not spread out like a blooming flower, or grow from a center, like continental ice. It grew all at once all over the plain. It filled in the mountain front. In the late winter of 1938, thirty inches of rain had fallen in six days. Publications of the United States Forest Service later described the event as "the greatest rainstorm in the recorded history of the San Gabriel Mountains." Early debris basins did not pass this test. They filled up and overflowed. They were not designed to separate rock from water. With the rise of population, and the postwar proliferation of development along the mountain front, successive debris dams and accompanying basins—each unique, fitting the conditions of its canyon—became ever more proficient, designed to function like giant colanders. The most recent is Buena Vista Debris Basin, which was built in 1981 at the head of Norumbega Drive, in Monrovia. With its Yale bowlishness, its columnar steel trash racks, its perforated outlet tower, and its two-hundred-foot-wide rock-filled concrete dam, it represents the state of the art. The confident owner of a house just below it has landscaped his place with sandbags.

Generally, the debris basins work. People can lie peacefully in their beds and listen to the thunk of boulders heading into traps. Carter Canyon Debris Basin, in Sierra Madre, is in an extremely steep gorge under high-rising mountainsides. Carter Canyon is where Wade Wells, one arid afternoon, pointed out to me a number of rock outcroppings about thirty feet above the dry streambed and said they were the lips of falls. The debris basin was built in 1954. In the autumn of 1978, eight different fires denuded the mountains above. The Mountain Trail Fire—the one most closely related to Carter

Canyon Debris Basin—consumed fifteen hundred acres. During a storm a few weeks later, Wells went to Carter Canyon and watched hundreds of tons of mud and boulders coming over one of the falls. The basin caught it all effectively. The basin was cleaned out, and trucks took away the debris. In 1980, when Carter Canyon Basin filled again, water alone went over the top.

There cannot be a debris basin under every rock and rill. Dennis and Susan McNamara live at 3529 El Lado Drive in Glendale, below the mouth of a minicanyon that drains about twenty-five very steep acres covered with chaparral. They had lived there for five quiet years when the watershed blazed like a ceremonial torch, September 22, 1975. Came the winter rains of 1976, and a debris slug appeared in their back yard. It filled their swimming pool. It continued across the patio and broke through glass panels into their house, where it eventually found the kitchen door and proceeded into the street. In 1977, another debris flow appeared in the McNamaras' back yard, filled the swimming pool, advanced into the house, found the kitchen door, and went out to the street. In 1978, when still another debris flow appeared in their back yard and filled their swimming pool, the McNamaras were nothing if not experienced. They slid open their glass doors and stood aside from the entering rocks. "Just the one fire started all this," Susan told me not long ago, as she showed me around. The by now revegetated slope was a steep handsome backdrop to the house. The pool was sparkling—a little debris basin in itself, all cleaned out and ready for what might come. "When it comes, it sounds like an avalanche," she said. "You can hear it start to rumble. The rocks crash against the house. Afterward, we thought of leaving, but you can't sell a house in this situation. You'd have to tell the buyers the endangerment of it. If it rains, we get very apprehensive. We still call each other and say, 'Are you all right?' "

The City of Glendale responded to the McNamaras' situation by giving them free grass seed. "That's all," Susan confirmed. "I can't say enough about the City of Glendale. In Palos Verdes—for the rich people—they *move hills*."

A debris basin that serves well in one year may not be adequate in another. It may be able to handle a fire-flood sequence of the magnitude for which it was designed, but a debris basin is not a strip mine. After a truly exceptional fire followed by a most exceptional flood, something like a strip mine is required. Designs are focussed on the "ten-year flood," the "twenty-five-year flood," the "fifty-year flood," and maps depict the "once-in-a-hundred-years flood," but these terms rest on data of only a century and a half and represent educated guesses. Rain, moreover, is merely one factor. No wonder there are times when the basins fail. When the debris hit Glendora in 1969, it scarcely slowed up as it filled two debris basins with an aggregate of a hundred and fifty thousand tons and flowed on into town. If you go up Country Club Drive, in Sunset Canyon, Burbank, you note a thick rind of defenses. With shored timbers, with six-foot walls of reinforced concrete or piled stone, properties are presented to the narrow street like medieval façades to an open sewer. There are three debris basins along Country Club Drive. There were two in 1964. The upper one failed. The slug that came down the street and invaded houses killed Aimee Miller, the wife of Frank Sinatra's piano accompanist. Her home was knocked off its foundation. Her husband was swept downhill and into a debris basin. He survived by hanging on to a Volkswagen that was part of the debris. One of their neighbors said, "When you live in a drainage ditch, you come to expect these things." Another said, "People often ask why we continue to live here. We have a fire nearly every year, and the floods follow. There isn't a prettier, more secluded canyon in Southern California—when it isn't on fire or being washed away. Each time we have a disaster, only one or two families move out, but there are hundreds standing in line waiting to move in. People live here, come hell or high water. Both come, and we still stay."

In the same era, a debris barrier failed above the ranch of G. Henry Stetson, the hatmaker, who had established his spread at the west end of the San Gabriels, in—of all places—the mouth of Sombrero Canyon. Below the pantiled roof, the balconies of the hacienda dripped with bougainvillea over mud deeply entering the doorways below. The floors were covered three feet deep. The furniture was destroyed. Lawns and gardens were buried under boulders and mud, which filled the Stetsons' million-gallon swimming pool and spread through their orchards, too. Sixteen feet above the streambed, a reinforced-concrete crib member that weighed two hundred and eighty pounds smashed against a tree, and the four reinforcing rods were bent around the trunk. In all, twenty-five thousand tons were added to the Stetson Ranch. The freshly burned watershed yielded altogether at least four times as much. The Stetson Ranch exists no longer. A new debris basin was built above the site. Its name is Sombrero.

Despite the recurrence of events in which the debris-basin system fails in its struggle to contain the falling mountains, people who live on the front line are for the most part calm and complacent. It appears that no amount of front-page or prime-time attention will ever prevent such people from masking out the problem.

Woman at her mailbox on a street called Bubbling Well.

In 1969, under six feet of debris, Bubbling Well stopped bubbling. "They've built more dams up there in the mountains," she says. "It will never happen again."

There are lots of mailboxes along the mountain front.

"There is no problem now."

"Our property is not affected."

"I think we're all pretty well covered. I'm hoping so. The water rushes down quite nicely in the wash."

"The fire was three years ago. It's not supposed to flood. I cancelled my flood insurance last year."

"No, we're not concerned. We lived through the San Fernando earthquake."

Many people regard the debris basins less as defenses than as assaults on nature. They are aesthetic disasters. To impose them on residential neighborhoods has been tantamount to creating a Greenwich full of gravel pits, rock quarries at either end of Sutton Place. The residents below Hook East were bitter when the basin was put in. Months later, the bulldozer tracks were still visible, they said, meaning that nothing had happened—no debris had come, and not even enough rain to obliterate the tracks. So why had the county used taxpayers' money to build something so obviously unnecessary? A form of answer came when the basin overfilled in one night. Afterward, people criticized the county for not building basins of adequate size.

When I talked to Peter Fay, who lives just below Sierra Madre Dam, he said there was a fire in his area every six or seven years. He found that "very exciting, almost fun," he said. "I have no intention of leaving the foothills. The views are so marvellous. The level of concern is quite low. If you live up in the hills, you have less smog, better views, more interesting contours—a sort of bosky sort of place. Part of my house was destroyed in 1969, but I am confident it won't happen again." Fay is a professor of history at Caltech.

In 1977, when Dan Davis was an engineer at Flood, he approached another homeowner below Sierra Madre Dam to warn him of the almost certain consequences of a recently extinguished fire. The man said he had lived through the 1938 flood, he had lived through the 1969 flood, and he was not concerned now. He must have had hide on his teeth. When the 1978 debris flow came, his place was buried deep in debris, but he survived.

People of the mountain front make remarkably hardy clients for consulting geologists and engineers:

"If you point out a potential geologic hazard, they say, 'I'm ready to live with that.' People really don't believe what it's like until they go through it."

"Most people don't give it a thought. Or they minimize the importance of it."

"People forget so soon. In a couple of years, they build again."

"They are playing ostrich."

As the larger rivers come into the Los Angeles plain, they sometimes thrash so violently that they pick up fresh debris from under people's houses. In 1969, when Marilyn Skates was twenty-six years old, she watched the houses of six neighbors go into Tujunga Wash. Her own house, swaying, hung on the brink. "What good does it do hanging there like that?" she cried. "Why doesn't it fall in like the rest? Go. Go. Go. Fall over. How can a house just stand there like that?"

It couldn't. The house slid into the Big Tujunga and was ground up so violently that pieces of it flew into the air. Her neighbor James Dubuque, out on the street and concerned about the fate of his own house, was about to go back inside when a police sergeant said through a bullhorn, "If you go back into that house, prepare to take your last step. That house is going into the wash—now!"

DAN DAVIS SHOWED ME some pictures of houses wrecked by a debris flow in Ebey Canyon that same year. I asked him, "What do people think when they buy such places?"

He said, "They never see the previous event. They move up there five, six, seven and more years, and nothing happens. They don't realize the tremendous change taking place in the watershed as it gets ready for another fire. Eventually, fire threatens them. Then flood. And they are on the Santa Anita Fault. Everywhere on the mountain front, people say, 'I've lived here x years. I've never seen any flow.' They won't, until there's a fire."

When I talked with Richard Crook, a consulting geologist, he said, "Most people are not conscious of the problem. Developers will buy a piece of property without thinking, for example, that it's at the mouth of a drainage. The engineers should be cognizant of it, but they're not. There are ordinances now, but in the past there were not. Every bad year, the agencies stiffen their rules. All this assumes that we know what the maximum situation is. Maybe we have not yet seen the hundred-year flood, or even the fifty-year flood. There are a lot of disasters up there waiting to happen. The people want to live in these areas. When they buy houses, they don't know what they are getting into. The entire county ends up paying for these people's problems. The people should be assessed for these things. They are localized problems. The whole county is subsidizing people on the front line."

Los Angeles Against the Mountains

Not to mention the federal government. Los Angeles is obviously rich (an indigent, threadbare city could never put up the defenses that Los Angeles has rigged against the attacking mountains), but no city is rich enough to pay off the family of every soldier who dies at the walls, or underwrite every house that fills up with boulders. Enter the generous people of the whole United States, whose revenues will help to cover the loss of a house constructed in the barrel of a canyon. The federal insurance is expensive. Its varying rates are based on zones of relative danger and on the quality of local flood-protection programs. On the average, a yearly premium of just under a thousand dollars will buy the maximum: two hundred and forty-four thousand dollars' worth of coverage. One does not have to be an actuary to see the general odds. Of, say, ten thousand houses along the mountain front, three to four hundred will be zapped per decade. A few private companies write such insurance as well. You can get lower rates if you have a "floodproofed house," which means that you are armed with deflection walls and other defenses to repel debris and send it elsewhere. In which case you also need a lawyer. Against debris flows and all other kinds of flood eighty per cent of the mountain-front people have no insurance at all.

It's an interesting bet. If you don't have a fire, you could go fifty years without a flood. After a fire occurs, people from the Sedimentation Section of the Hydraulic Division of the Los Angeles County Department of Public Works are swift to tell homeowners what to do when rain comes. After studying someone's property, they present a "Post-Fire Protective Advice" sheet, on which one or more of twelve printed suggestions have been checked off. One suggestion is "Construct pipe-and-timber deflector." Another is "Board windows and/or doors with plywood." Another is "Consider evacuation during heavy storms."

Arthur Cook, the Acting City Manager of Glendora, recalls with some choler the procedures of 1968 after twenty thousand acres burned above town. When I visited his office in City Hall, he said, "The whole face of the hills burned. We started right then to prepare for the winter rains. We knew what would happen. It was evident. We went to the government agencies—county, state, and federal. We were met with total negativism. They could not help *prevent* a disaster, they said—they could only come in after the fact. We thought, If no one's going to help us, we'll try to help ourselves. We held neighborhood meetings. We recommended plywood walls, block walls. We told people they'd have to make their own preparations. People said we were alarmists. They'd lived here fifty years, and, by God, it had never happened. So the city installed batten boards along some streets. On Rainbow Drive, we erected chicken-wire fences with sheets of plywood. We deployed thousands of sandbags."

Sally Rand lived in Glendora, pretty far up on the fan. At Christmastime that year, in the Kingdom of Rubelia, she did her phan dance at Mike Rubel's castle. Four weeks later came the devastative rain. Listening to it thump his roof, Art Cook said to himself, "This is the night. I'd better get down to the Hall." He got into his car and went down Palm to Grand; as he turned onto Grand, the flow met him. It picked up his car and carried it a quarter of a mile. ("I just sat there, scared as hell.") Eventually, his slithering wheels found traction, and he drove off none the worse for having been a part of the snout of the debris. "A wave six to eight feet high came out of Rainbow Canyon," he said. "The rock, debris—everything was suspended in the liquid mass. Horrendous boulders, trees, car bodies were suspended in the mass. It sounded like a train—a runaway express train. Just a roar."

As an average populace in a shamefully litigious nation, the people of Glendora took their town to court. "After we busted our butts for months trying to help people, we ended up being sued by the people we tried to help," Cook said. "They said in court that we were cognizant there were going to be problems and we did not stop Mother Nature in her tracks. You read about flooding on the Mississippi. They don't sue the city for not keeping the Mississippi from going over its banks. Here they do sue the city for not keeping the mountains from eroding." The city lost.

I asked Cook how things appeared at the moment. After mentioning a couple of new debris basins and some crib dams, he said, "The areas are protected now. People rebuilt. People had their home and they were going to keep it. People started listening to City Hall. There's no way to guarantee that Mother Nature won't go on a rampage, but if it did we have the facilities to cope with it. Now we're in great shape, with no concern for the future."

In the fifty miles of the San Gabriel front, almost all levels of income are represented, from multifamily dingbats to pastury enclaves with moats and armed guards. Under isolated tufts of old citrus are electrically operated iron gates leading to palm-lined driveways between retaining walls of granodiorite debris. Relentlessly, builders go on finding tracts for new housing. They often borrow something from the precipitous slopes—a practice known as mountain-cropping. Whether an owner has a two-million-dollar house protected by nothing more than a trash rack in front of a culvert or a hundred-and-fifty-thousand-dollar hut with a large debris basin right beside it, debris-flow information is supposed to be a part of the exchange when houses are sold. For example, here's a nice little bungalow for sale pretty far up the fan in the neighborhood where Art Cook and his automobile were given a quarter-mile ride by the debris flow of 1969. Three bedrooms. A quarter of an acre. How about debris flows?

"Not that I know of," the realtor says. "I've never heard of that happening up there."

House, swimming pool, Jacuzzi, forty yards from Pine Cone Road, $274,500. Could there be a debris problem?

Crescenta Valley Realty: "No. There is not. There is not any problem."

"Why not?"

"It is well maintained."

Pasadena Glen. Two bedrooms, one bath, nine hundred and fifty-two square feet, $129,500. Flood and mud?

Realtor: "There has been in the past—some flooding in the lower areas. But this one is up high. We have to make buyers aware that they are in a special-study zone."

A special-study zone is a place where debris is likely to interrupt the special study, where canyons aim at homes, where insurance rates are high.

New home, built by developer, on one acre, on Ben Lomond Drive, Bradbury. Four bedrooms, four baths, three thousand square feet, $337,500. Flood and mud problems?

Coldwell Banker: "Not to my knowledge. They would have to be disclosed, too. Each listing has a seller's disclosure form, and there's nothing about mudflows on the form."

House on Bubbling Well, Glendora, $167,000. Mud and flood?

Alosta Realty: "No, no. Oh, years and years ago we had a big flood. They built that dam, and there's no problem there at all."

Silent Ranch Estates, shake-roofed or tile-covered castles, "individually built," in the four-hundred-thousand-to-one-million-dollar range. Possible debris-flow problems?

Realtor: "There are a couple of houses which I would not touch. The others—I don't think so. I haven't heard anything to that effect. Every house you'd see, we'd have to take that into account."

Terrace View Estates, gated community, all view lots, up to one and a half acres at roughly $135,000 an acre. Debris flows?

"No. We went through the whole winter there. We had some very heavy rains and there were no problems."

New development in a former orange orchard—top of the fan. House for sale: four bedrooms, exercise room, sunken shower, sunken tub, elevator, $1,100,000. Flood and mud?

Realtor: "I have watched this project for over two years. There is no evidence yet of anything. Our company bought three lots there. We would not have invested in these places if there was a problem."

The house across the street has forty-five hundred square feet of living space, not including its three-car garage; a dry gully in the back yard is spanned by a footbridge—$695,000. Although the street is a cul-de-sac, it has evidently been designed as a conduit for more than motor vehicles. The houses beside it are high on built-up banks. Below the turnaround at the bottom is the mouth of a concrete channel, waiting to take in whatever may come. Above the tract, a developer-built debris basin and trash rack call to mind a teacup and tea strainer. The slope rises through live oaks into soft chaparral.

Three-bedroom tri-level, two and a half baths, on two-tenths of an acre in Sunland, just inside the Los Angeles city line. Den, wet bar, pantile roof. $245,000. Mudflow problem?

"I have worked in that area since 1979, and there has not been, to my knowledge, anything large up there."

Flat roof, carport, thick-picket fence, fifth of an acre in Sunland, just inside the Los Angeles city line. Debris?

"Trust me. I have lived many years in hill communities. About ten years ago, there *was* a problem. It had to do with the way they had graded the cemetery."

Six bedrooms, septic tank, three baths, fifth of an acre in Sunland, just inside the Los Angeles city line. Mud-flow problem?

"Yes, it's always a problem. Special insurance is necessary."

Two bedrooms, one bath, $91,000, in Sunland, just inside the Los Angeles city line. The house is on Hillrose Street directly below Verdugo Hills Cemetery. Could there possibly be a debris-flow problem?

"No. Not at all."

Four bedrooms, three baths, pool, third of an acre, $239,900, on Hillhaven Avenue, Sunland, just inside the Los Angeles city line. Is there a debris problem?

"If you make an offer, the seller must tell you. A cemetery slid out a couple of years ago."

Other houses, other realtors, same neighborhood:

"I don't know. I don't think they have that problem there."

"There was a problem years ago. They have a channel there now."

"It's always possible, but that place is not on enough of a hill."

"Not that I know of."

"That I don't know. You have to check with the City of Los Angeles."

CARL GUNN CAME TO Los Angeles with his parents in
1916, "when you could drill a hole ninety feet deep and the
water would flow out and irrigate your eighty acres." Born in
Iowa and made of baling wire, he has been in Glendora fifty
years. Like Chuck Colver, he antedates the debris basins and
the people they protect. He still lives beyond all defense, far
up the bed of a canyon, in what is locally called the elfin
forest. His house is small, one-story, with tractors, disk har-
rows, and other machinery around it, at the base of a slope
that rises fifteen hundred feet in half a mile. Retired now, and
close by a wood stove, he used to work the Blue Bird Ranch.
He leased four hundred and forty acres and raised cattle and
lemons. He said, "When I was farming the Blue Bird and we
got forty inches, we was tickled to death—cattle running
around in grass that was up to their belly. There were some
beautiful silt valleys here. Now it's all houses and streets. The
old ranchers and farmers are gone. Damn few know the story.
They're newcomers. Ninety per cent of them never know that
the water can go down there like the milltails of Hell. They
have no perspective on the possibilities. The people who buy
the houses don't know that sooner or later stuff is going to
come down through here like shit through a tin horn." Gunn
once built himself a crude two-room cabin. It recently changed
hands for two hundred and fifty thousand dollars. In the canyon where he lives now, a loud sound woke him up one winter
morning at six o'clock. It was the sound of errant debris breaking into his house. "Before you could say 'scat,' the water and
mud were three feet deep. I didn't hardly believe it. It doesn't
bother me. I know that it happened. If a fire comes, then I
worry. I wouldn't trade this place for a dozen others, even
though you know what can happen. It's the privacy. The

coyotes. The deer. If it gets where I can't pee off my front porch, I'll move."

Chakib Sambar, who was born in Lebanon and is now the vice-principal of Crescenta Valley High School, has lived against the mountains nearly thirty years. He has seen his share of debris—especially outside his back door, where it has built up in considerable volume in a crease called Eagle Canyon. When I called on him one day in early June, he handed me a 3-iron. As we began a short hike into the canyon, he explained that the sheriff had circulated a rattlesnake alert. "I've never been out here at this time of year," he went on. "I only come out in the spring and the fall."

I said, "You call this summer?"

He said, "When it hits a hundred, it's summer."

The canyon was loaded with debris. We walked through pale-yellow yucca. "This is California chaparral," he said. "When a fire hits this plant, it explodes. It actually explodes." Sambar's son Tod was with us. He did not carry a golf club and was apparently relying on his youthful agility to frustrate the baking serpents. Sambar carried a putter. "We have seen a lot of fires and floods," he said. "The original channel in this canyon was forty feet lower than it is now. This is the beginning of another disaster—a great mass of debris to go out in a storm."

"It seems to be aimed at your house," I said.

"Yes," he said. "It is." And he laughed as he added, "My wife is not very thrilled."

"Then why do you live here?"

"Actually, we moved up here because it was the only place we could afford. We bought our house for thirty thousand dollars. It is now worth ten times that."

When the Richter scale is busy, waves form in the Sambar

swimming pool. When fires close in, the Sambars have stood
on their roof to watch the flames.

"This canyon and this entire mountain were on fire for
forty days," he said. "You learn to adapt. You live with it.
Fortunately, trouble doesn't come too often. In the years we
have been here, we have had only one major earthquake, two
major fires, and one major flood."

In Los Angeles, a longtime resident born in Iowa or in
Lebanon is something like a royal palm. Neither is native and
both are common. Rare, though, is Miner Harkness. He was
born in Sierra Madre, and half a century later he still lives
there—right up against the hard chaparral, with a watershed
behind him that has an average grade of eighty-five per cent
and in many places is vertical. This is the country under Mt.
Wilson, the roughest segment of the range. Harkness has an
insurance brokerage, which is half a mile from his house. On
a wall there is an Ansel Adams poster called "Examples." In
the photograph, huge boulders lie at the foot of mountains.

Harkness likes to go up from town and into the loose rock.
"My big love is backpacking and taking day shots out and doing
peaks," he said. "I'm not a big-wall rock climber." The remarks
were made in his Toyota 4 × 4 while he showed me around
the nearby mountains. He was a small, compact man with
alert blue eyes and dark brows, an intelligent and kindly face.
His hair, generally thinning, was gone in the back, his beard
Vandykish and in two shades of gray. We were looking north
up Big Santa Anita Canyon—a lovely vista over crisscrossing
spurs. On a high mountain face several miles away was a region
that appeared to have been bared, and bared again, by the sort
of rock avalanche that comes away like a broken tooth and
exposes the interior of hundreds of acres of mountain. There
were dozens such. The denudation had been caused entirely

by human beings, and was the result of the switchbacks in a complexly engineered route built to enable trucks to get around a falls. "It's still eroding," Harkness said. "It's a permanent scar. We call it the Burma Road."

The trucks hauled the concrete bars and other components of some fifty crib structures that were emplaced in tributaries of Big Santa Anita in the early nineteen-sixties. Harkness referred to them as check dams, and explained—as Barclay Kamb had—that they were meant to prevent the streams from downcutting and thus reduce the watersheds' production of debris. Harkness said, "Big Santa Anita Canyon was a wild place with a beautiful trail in it. When they bulldozed that giant road, we thought they were *creating* more debris flow than they could ever hope to stop. With the check dams, they were virtually destroying the canyons. All the natural beauty was being virtually wiped out. I felt that what they were doing was wrong. It was costly to taxpayers, too. We're talking about millions and millions of dollars, wasted. There is absolutely no good reason to put those check dams in. They were doing this in canyon after canyon after canyon."

It was an era when even the newspapers were reflecting the confidence of the county engineers—when, according to one reporter, "stabilizing the steep banks will, once and for all, put an end to erosion."

PROJECT AIMS TO HALT
EROSION OF MOUNTAINS

VALLEY AUTHORITIES VOTE
LANDSLIDES UNNECESSARY

"Flood Control and the Army Corps of Engineers have lost all perspective as to what they're all about," Harkness said.

"The difficult thing is, they get paid, and we—who just live here—don't. If we stop them, they come back a few years later and do it again. They refer to flood control and saving human lives—they push it on that basis—and that's a lot of bullshit. They were going to do Little Santa Anita, cutting roads in six or seven miles. Can you imagine the debris flows that would be caused by *that*?"

The check-damming of Little Santa Anita was stopped by Miner Harkness, Barclay Kamb, Peter Fay, and others, whose organized efforts dissuaded the Los Angeles County Flood Control District and the Los Angeles County Board of Supervisors. While the battle was going on, a volunteer who spent his evenings knocking on doors to enlist opposition to the check dams spent his days in the mountains as a heavy-equipment operator vigorously making check dams.

"We were not just the Sierra Club raising hell," Harkness said. "I have great respect for the Sierra Club, but I think at times they oppose everything that comes along. I'm not just some nature nut. I've been in the mountains, and I know what the mountains are about."

Harkness has seen three mountain lions—creatures so elusive that a person can spend a lifetime trying to see one. As a kid, he hunted deer with arrows. When the red cars of the electric railway came up to an earlier Sierra Madre, people got off and hiked into the canyons. At night, they sometimes carried tin cans cut through on a slant to hold and reflect lighted candles. They walked up Little Santa Anita Canyon and on to the observatory on Mt. Wilson, climbing four thousand feet holding candles. In the mountains above Sierra Madre lived four Lalone brothers, who trapped coyotes and cut wood. Unsurprisingly, they were often called upon to search for and rescue people whose candles—this way or that—

went out. The efforts of the Lalone brothers evolved into a group called Sierra Madre Search & Rescue. The team is twenty-five in number now, and Miner Harkness has been a part of it for thirty-five years.

We drove downstream toward the city. Near Big Santa Anita Dam, which was built in 1927 to impound water, he said, "Kids climb the walls of this beautiful canyon. They get scared and freeze. The rock is decomposed granite, which is a climber's nightmare. In the Sierra, the granite would hold a two-ton truck; here it won't even hold my hands pulling down on it. We rappel to the kids. Tie them off. Get a rope and helmet on them, and bring them down."

These were the first people to train bloodhounds for mountain rescue. They were the first to rappel from the air. On a rescue in Bear Canyon, ten miles from home, Harkness was in a helicopter and, in his words, "there was just no place to land," so he stood on the skids and jumped to the ground. That was the impulsive preamble to a rappelling procedure now known as helitak, which is practiced by mountain-rescue teams everywhere in the world.

The members of Sierra Madre Search & Rescue are so skilled and so famous that they have been called to emergencies in, among other places, the Adirondacks, Iowa, and Mexico. Needless to say, they cover the whole of California. Their phone number is on a wall at Yosemite National Park. They conjoin sometimes with the United States Air Force—C-130s with Sierra Madre trucks inside. In the nineteen-seventies, they thought of lowering bloodhounds from choppers in slings, easing them down into the chaparral. The procedure has its limitations. A lot of San Gabriel terrain is much too rugged for dogs.

In Sierra Madre's City Hall, the team has a den they call

the Rescue Room. Citations adorn the walls—from the American Red Cross, from El Gobierno del Estado de Baja California, from the Department of State of the United States of America. Hanging from the ceiling is a helicopter's tail rotor bent like a boomerang. A parachute is up there, too. Now that they can rappel from helicopters hovering as much as a hundred and fifty feet off the ground, they no longer have a need for parachutes. The Rescue Room is furnished with seats from a 747.

As a consequence of jumping off helicopters and carrying heavy backpacks, Miner Harkness walks with a very slight limp. On New Year's Day, he rides. When the community of Sierra Madre constructed a Rose Parade float to display the Search & Rescue team, the float was a rolling mountain.

At the upper edge of his home town, we went into Sierra Madre Canyon—a former summer colony, a narrow wash of the Little Santa Anita River. It was something like Pasadena Glen, a couple of miles away, but instead of Gunite streambanks—not to mention the reinforced-concrete box channels prevalent elsewhere—this unaltered wash was full of stream-rounded cobbles and brought to mind villages of the Pyrenees and villages of the Massif Central. "Some houses in this canyon you walk up a trail to get to," Harkness said. Most were extremely close to the wash and to each other. There was a covered bridge over the wash. At the head of the street was Sierra Madre Dam, larger than an ordinary debris basin but modest as a dam. Harkness said he had seen logs three feet in diameter shooting over the spillway like canoes. Los Angeles County and the U.S. Army Corps of Engineers had proposed expanding the dam and building crib structures far up the canyon the better to arrest the debris. "They spoke of major fires, of the hundred-year flood. It's always the same tactics.

But the people of Sierra Madre said, 'Buzz off!' We said, 'Hey! We love our canyon as it is. We'll take the risk.' They used constant scare tactics, saying that the dam was cracked, and so forth. Then they said, 'If you people don't accept our recommendations, you people are going to be legally liable for the City of Arcadia.' " Arcadia lies below Sierra Madre, at the bottom of the fan.

"They go into some town and the people are sold this bill of goods," Miner Harkness went on. "The next thing you know, it's done, and the people say, 'Jesus Christ! What happened to our canyon?' That will not happen in Sierra Madre. The people love the canyon, and they get up in arms. In Sierra Madre when there's a fire, the whole town comes up to the mountain frontier to stand on roofs and soak them down. That is the spirit that keeps the Flood Control people and the Army Corps of Engineers from tearing this canyon to shreds."

He took me up to high ground, where we looked out into a fading haze at a large part of Los Angeles. The fan, below us, was a paisley of swimming pools. There were million-dollar houses on some truncated hills. Harkness said, "People who come up here and knock off the tops of these mountains— if they get wiped out in a debris flow or a forest fire, that is the price they pay for the view. The county or the state should not have to foot any of that. In a person's tax bill, though, a high percentage of the total is for flood control. That's wrong. Living there is a risk people should be willing to take. A major brushfire declared a national disaster! That's b.s.! I know the risk in building a house there. You build there or buy there because you want to be there. Other people shouldn't be required to pay for my risk."

One February morning years ago, Miner Harkness stopped downtown for a quick cup of coffee before flying a mission in a Los Angeles County helicopter. Some firemen were drinking coffee, too. A fireman's pager sounded. A debris flow had mobilized, and people were trapped.

Harkness felt "a real cold chill." Of the several dozen most vulnerable houses in Sierra Madre, his was one. He went home immediately. All around the place, red lights were flashing.

His house had been torn in two, with his wife, Sara, in it, and their infant daughter, and their five-year-old twin sons, another daughter, and Joanne Crowder, a next-door neighbor. In a small canyon above, a large amount of debris had bulked up through time. The construction of a fire road had contributed to the volume. As people so often do, Harkness compared it to a loaded gun. Sara and the baby were in the master bedroom. When the slug of mud and boulders came out of the mountain crease, Joanne Crowder, who had fled to the Harkness house, was caught by the entering debris. It broke her back and a collarbone. The Harknesses' hot-water heater was flung upon her, and the water scalded her legs to the third degree.

The Harkness house projected from the hillside and had a carport beneath the master bedroom. The debris tore off the master bedroom with Sara and the baby inside. The bedroom fell on the family station wagon. With the bedroom on top of it, the station wagon went down the driveway and on down the street. In what remained of the house, the twins and their sister Claudine were unhurt. Sara and the baby came to the end of their ride unhurt. The station wagon suffered considerably. When the bedroom was taken off it, the car was twenty-six inches high.

The living room had been knocked four feet askew but had not collapsed. A bathtub was ripped nearly in half. Almost surely, the master bedroom would have broken into pieces if it had not fallen on the station wagon and been carried away intact. Furniture and jewelry were all over the hill—"heirlooms everywhere, but there was no looting; no one took a thing."

The baby, Andrea, slept on her father's chest for six months after the event. She still gets nervous in a lightning storm.

Naturally, I asked Miner Harkness if he had contemplated moving away.

The look that came into his face suggested that he was confronting a strange, unprecedented question. He said, "It never entered my mind." He remembered Sierra Madre when it was an isolated town. He had seen the subdivisions change the country where he hunted as a boy. He had built his house as close as he could to his predilected mountains. It was on Churchill Road off Mountain Trail, just west of Sierra Madre Canyon, on a site where there had once been a hunting lodge—for the people from Los Angeles who came up to Sierra Madre on the railway with the red cars. Miner and Sara Harkness (an intensive-care nurse) were not about to move away. Their destroyed house was uninsured. Inspection showed that not much could be kept but a small block wall. Engineers who looked over the site told the Harknesses not to rebuild. When Miner made clear to them that such advice would go unheeded, they suggested a monumental retaining wall—a sort of one-family check dam—to try to stabilize future debris.

Building a check dam was not his style. "The heavy equipment would tear up the canyon," he said. His solution was a house enlofted. It stands in the air, on sixteen concrete pylons, each fifty-seven inches around. The pylons go down into the earth more than ten feet. Inside them are bundles of reinforcing rods wrapped in sheaths of steel.

The house itself is sheathed in cedar, and has an outside stairway of rough-hewn planks. Its windows are large and without curtains, which are not needed among the evergreen oaks. When the city emerges from beneath its veils, nothing impedes the view. The place has the aspect of a lookout tower positioned on the front—a post of observation in the cycle of fire and rain. Seven times, Miner Harkness and his family have been ordered to evacuate their house because of the threat of fire. And after the fires the dry ravel has built new depths of debris. Harkness clearly sees himself as an indigenous part of the cycle. Sitting on one of his rough planks, he said, "When the Santa Ana winds blow, and a big fire gets under way, you're not going to be able to hold it if it ever comes through here. That is a risk you take."

IF LOS ANGELES hangs on long enough, it will cart the mountains entirely away, but already it is having difficulty figuring out where to put them. In a productive season, the debris basins will catch more than a million cubic yards. The reservoirs back in the mountains take in much more than that. Over-all cleanout costs can exceed sixty million dollars in a single year—a required expenditure if the system is to function.

The Los Angeles side of this battle is unexampled in heroic chutzpah—to a degree expressible in the volume of any fan. For a brief geologic time, debris has been pouring out of Little Santa Anita Canyon, for example, and piling up on the plain. Among the San Gabriels' alluvial fans, this is a modest one in all respects. Its toe is down at the horse track—Santa

Anita—and its slope begins to rise through the Los Angeles Arboretum and then goes on rising a couple of miles, through the center of Sierra Madre and on up Baldwin Avenue to Carter Debris Basin, at the mountain front. The fan coalesces with fans on either side, but its width can be approximated at a mile and a half. Calculation shows that this one minor fan consists of about two billion cubic yards of material that has so far come off the watersheds above. Toward the middle of the twentieth century, Los Angeles undertook to replace nature as the depositor of material. Los Angeles assumed responsibility as well for at least fifty other fans on the San Gabriel front, not to mention lesser uplifts, like the Santa Monica Mountains and the Verdugo Hills.

Against the prodigious odds, Los Angeles is much more successful than it may appear to be when neighborhoods buried to the rooflines are pictured in the *Times*. To date, the debris basins have trapped twenty million tons of mountain. A small fraction of that figure has managed to get beyond them. Since the basins can fill up in a few hours, equipment has to be marshalled very swiftly to make cleanouts—and sometimes repeated cleanouts where more debris is expected. At the height of a storm, radio patrols are all over the front feeding information into the computers of the Storm Operation Center, which deploys forces from six field yards—deploys the motor graders, the backhoes, the front-end loaders, the trailer-mounted night-construction floodlight systems, the thirty-five-ton truck-mounted clamshell cranes, the forty-five-ton drag-lines with six-yard buckets, the six-inch pumps that can suck in and spit out rocks, the rock grapples that can pick up five-ton boulders, and, above all, the big ten-wheelers: dump trucks capable of hauling fifteen tons at a time.

Los Angeles owns some of this equipment, but Los Angeles is not an OPEC country. Los Angeles cannot afford to keep hundreds of dump trucks waiting for an annual or biennial storm. The debris is made attractive to private truckers. They are paid fifty dollars an hour for lugging mud, no overtime. Flood has used as many as three hundred private trucks in one storm. They come from all over Southern California—from Lone Pine in the Owens Valley, from San Diego. Forty trucks once came from Redding, nearly five hundred miles north, and found no work when they arrived. A call once came to the section from a heavy-equipment operator in the Midwest. He wanted the work for his cranes and large loaders, and was willing to send them two thousand miles.

Wherever there has been an antecedent fire, debris basins are likely to fill to the brim. In 1978, every basin filled for fifteen miles under the slopes burned by the Mill Fire—eighteen basins in all. In some of them, the rocks were so big that they had to be broken by dynamite before they could be removed. Dunsmuir was cleaned twice that winter, Mullally and Denivelle three times. They filled as they were being emptied. Zachau was cleaned out three times as well, but it still let boulders ten feet in diameter get away. People living below Zachau sued (unsuccessfully), claiming that the county had not maintained the basin. "We had records that showed if there was two ounces of dirt in it we took it out," Dan Davis says, his indignation unmitigated.

Before the creation of the debris basins, mountain sands were carried to the ocean by winter floods. Now there is a beach problem. Sand is being lost to offshore canyons and is not being naturally replenished. A place that values its beaches as much as Southern California does has no choice but to buy sand. Sand has been transported by ten-wheelers from the mountains to the beaches. When thirty thousand yards of sand was put on Zuma Beach, people complained about the color contrast. Materials cleaned out of the Laurel Ridge Debris Basin, in the Santa Monica Mountains, used to be hauled over the mountains in a direction away from the ocean and dropped at the Calabasas dump. Subsequent Laurel cleanouts went onto the beach. The Sedimentation Section has been investigating the possibility of using pipeline slurries to transport debris to distant gravel pits; it could also go in pipelines to the beach. Vito Vanoni said, "I think the day will come when we grind it all up and send it to the beach. The question is: Where do we get the water for the slurry? We could use sewage. That's not so good. Possibly we could use salt water."

Years ago, in an act of lyric irony, Flood Control bought four cloud-seeding generators and set them up near reservoirs in the mountains and along the front. The cloud-seeding generators have been used almost exclusively during foul winter weather, since they would not be efficacious under the otherwise azure sky. The department is criticized for seeding clouds. On the other hand, letters appear in the *Times* attacking the department when it does *not* seed clouds. The generators shoot incoming weather fronts with microscopic crystals of silver iodide. This is known as "enhancing the storm." The storm is worth at least a hundred dollars an acre-foot. Los Angeles wants the water so much that it mines the storm. This requires artful judgment. The idea is to increase the volume of rain, but not to the point of mobilizing debris flows. The cloud-seeding generators were running on February 7, 1978, for example, and they ran on February 8 and February 9, but they were shut down when the proportions of the storm became apparent. The events on Pine Cone Road and at Hidden Springs and in the Verdugo Hills Cemetery, among others, occurred in the first hours of February 10. The seeding could not have added anything much to the total rainfall, but the fact remains that during the wettest week of the rainiest season of the twentieth century the cloud-seeding generators were enhancing the storm. Cheeky is the warrior who goes behind the lines to pick the enemy's pockets on the eve of battle.

In 1980, when John Tettemer was Acting Chief Deputy Engineer at Flood, he remarked at a symposium, "The war stories we tell each other are almost exclusively related to debris problems. . . . We are all getting blitzed after fires by small canyons. . . . Funny things are occurring in small debris basins. I cannot tell you exactly what. All I know is that during major storms debris sometimes goes over the spillway when the basin is not really full. We do not understand the dynamics, but it seems that these small basins act more like flip buckets than lakes. . . . We and the U.S. Army Corps of Engineers expected our channels to last maybe a hundred years. We have seen cases where the reinforcing steel was exposed in one storm. . . . We do not know enough. . . . We also find communities starting to wonder why their high-priced facilities fail during storms when they are needed. I do not blame them, do you? . . . We should stop building things where they do not belong, and leave some room for nature." To which his colleague Arthur Bruington added, "Through the district, the residents are battling, but sediment is still winning. . . . Managing the sediment capable of being produced by the San Gabriel Mountains is important to the safety of the residents of the district, but, frankly, it is like trying to hold back the storm tides of the ocean."

After one season, twelve million cubic yards were removed from four of the mountain reservoirs. After one season, a million two hundred thousand yards were taken out of the debris basins. In Santa Fe Basin, the huge trap in the San Gabriel River where it meets the Los Angeles plain, the Army Corps of Engineers has fifteen million yards it wants to get rid of, but where? Disposal sites have become extremely hard to find. Every abandoned gravel pit between Tujunga and Twenty-nine Palms has already been filled with the rock of the transverse mountains, or so it seems. Chuck Colver, of Covina, who watched his orchards go down and his neighbors come up, has been around long enough to see a likely solution to the problem. "One of these days," he predicts, "they'll be buying tracts of houses to get a place to deposit the material."

Donald Nichols, of the Department of Public Works, pointed out to me one day that the word "debris" suggests sanitary landfill and odor, so debris-disposal sites are now called "sediment-placement sites" in order to get communities to accept them. "To have to stroke our folks like that kind of insults me, frankly," he said. "But disposal is a real dilemma. The Sierra Club says, 'If you put it here, you'll kill a lizard.' The Forest Service and Fish and Game also complain. Acquisition of debris-disposal areas is much slower today because of the California Environmental Quality Act. When we try to establish a new disposal area, we run into opposition from homeowners and environmentalists. The environmentalists run the gamut from the sensible ones to the crazies. They say, 'You'll kill what you put it on. You'll dislocate animals. You'll alter the land form.' That's correct. But when we're finished we'll plant vegetation. The animals will come back. The rest we have to leave to the Big Guy in the Sky, who will finally naturalize these deposits we make."

The Big Guy doesn't always do a perfect job. In Laguna

Canyon years ago, some forty houses were built on old disposed debris. In 1978, the fill failed. The forty houses—worth two hundred and fifty to five hundred thousand dollars each—slid downhill like sleds.

Inevitably, someone was inspired to put the rock back up in the mountains. This elegant absurdity may be the *ne plus ultra* in telling the Big Guy who's in charge. San Gabriel Dam, a few miles upriver from Azusa, was built in the late nineteen-thirties to keep debris from clogging a reservoir just below it. More than twelve million tons of debris have been stopped behind San Gabriel Dam in one rainy season. San Gabriel catches so much mountain that it has to be more or less continuously cleaned out. Fifty or sixty trucks have lined up to lug the debris away. One place they have put it is in the tributary Burro Canyon, a haul of less than a mile, and—in elevation—six hundred feet up. "That way the engineers can have job security," Wade Wells once said to me. "They take the debris and carry it back into the mountains, where they create a potential debris flow."

"Burro will someday serve as a campground," Don Nichols said. "We have improved on nature by putting a mountain up there that doesn't come back down. Burro has debris basins in it. It has *its own* debris basins. We put fourteen million cubic yards in Burro Canyon."

Wells and I went up there one day to see this epic artifact—in clear dry air and vast silence—eight miles from Arby's. A California quail ran by, sporting its knightlike plume. In the V-shaped mountain valley, the deposit rested like an aircraft carrier in dry dock—a comparison that would be more apt if aircraft carriers were not so small. Debris basins were along its upper flank, there to *protect* the man-made deposit.

Burro Creek passed under it, through a deep culvert a mile long. For twenty million dollars, Los Angeles had returned the rock to the mountains. For twenty million dollars, they had built in Burro Canyon an edifice ten times as large as the largest pyramid at Giza.